From Crisis to Stability: Understanding Energy Security in a Volatile World

Olugbenga Onungbogbo

OLUGBENGA ONUNGBOGBO

From Crisis to Stability understanding Energy security in a volatile world
Copyright © 2024 by Gbenga Onungbogbo

All rights reserved. No part of this book may be reproduced, stored in a retrieval system, or transmitted in any form or by any means, electronic, mechanical, photocopying, recording, or otherwise, without the prior written permission of the publisher, except in the case of brief quotations embodied in critical articles and reviews.

ISBN: 9798345744017

Imprint: independently

Printed in the United States of America

TABLE OF CONTENTS

Dedication

Acknowledgement

Preface

Chapter 1: Introduction

The Historical Evolution of Global Energy System

The Transition to Fossil Fuels: Coal and the Industrial Revolution

The Birth of Renewable Energy: Solar, Wind, and Beyond

The Energy Landscape Today: A Complex and Interconnected World

Chapter 2: The Dimensions of Energy Security

The Dimensions of Energy Security, Availability, affordability, accessibility, and sustainability.

The Interconnectedness of Energy Security Dimensions

Chapter3: The Geopolitics of Energy Security

The Global Energy Market and Economic Interdependence

Energy Security and Global Governance

Chapter4: The Role of Technology in Enhancing Energy Security

Emerging Renewable Technology

Emerging Storage Technologies

Carbon Capture, Utilization, and Storage (CCUS)

Chapter 5: Geopolitical Implications of Energy Security

Energy Resources as Instruments of Power

Energy Security in Trade and Diplomacy

Climate Change and Energy Diplomacy

The Role of Emerging Economies in Energy Geopolitics

The Shifting Balance of Power

Chapter 6: Navigating Energy Security and Environmental Sustainability

The Environmental Impact of Traditional Energy Sources

The Role of Smart Technologies,

Policy and Regulatory Frameworks

Successes and Impact of International Institutions

Challenges and Limitations of International Institutions

Case Studies of Successful International Initiatives

Recommendations for Enhancing the Role of International Institutions

Chapter 7 Power and politics: The Geopolitical stakes of Global Energy Security

Geopolitical Risks Associated with Energy Security

The Impact of Energy Security on Regional Stability

The Role of Emerging Economies in Global Energy Geopolitics

Strategies for Enhancing Energy Security and Managing Geopolitical Risks

Future Trends and Challenges in Energy Geopolitics

Evolving Geopolitical Dynamics

Chapter 8: Energy Security and Environmental Sustainability

OLUGBENGA ONUNGBOGBO

The Interplay Between Energy Security and Environmental Sustainability

Technological Innovations for Sustainable Energy Security

Case Studies: Balancing Energy Security and Sustainability

Chapter 9 The Economic Implications of Energy Security

The Impact of Energy Market Dynamics

Geopolitical Risks and Economic Stability

Policy Responses and Economic Strategies

The Role of International Cooperation

Future Trends and Economic Implications

Chapter 10: Energy Security in a Changing Global Landscape

Technological Advances and Their Impact on Energy Security

Economic Implications of Energy Security

Policy and Regulatory Responses

Environmental Considerations and Energy Security

Future Directions and Strategic Responses

Chapter 11: Energy Security and the Role of International Institutions

Overview of International Institutions in Energy Security

Successes and Impact of International Institutions

Challenges and Limitations of International Institutions

Emerging Trends and Future Directions

Case Studies of Successful International Initiatives

Chapter 12 The Role of Emerging Economies in Shaping Global Energy Security

The Growing Influence of Emerging Economies

Opportunities for Emerging Economies

Case Studies of Emerging Economies Shaping Energy Security

Strategic Recommendations for Emerging Economies

Chapter 13 The Path Forward: Strategies for Enhancing Global Energy Security

Strengthening Energy Infrastructure

Advancing Technological Innovation

Fostering International Cooperation

Implementing Effective Policies and Regulations

Addressing Geopolitical and Economic Challenges

Engaging with Stakeholders and the Public

Chapter 14: Navigating Energy Security in a Changing Climate: Adaptation And Resilience Strategies

Understanding Climate Change Impacts on Energy Security

Strategies for Adapting Energy Infrastructure

Innovations for Climate Adaptation and Resilience

Recommendations for Enhancing Climate Resilience in Energy System

Chapter 15 The Path Forward: Shaping a Resilient and Sustainable Energy Future

Embracing a Holistic Approach to Energy Security

Advancing Technological Innovation and Deployment

- Enhancing Policy and Regulatory Frameworks
- Promoting Resilience and Adaptation
- Ensuring Equitable Access and Social Considerations
- Case Studies of Successful Strategies and Innovations
- The Path Forward: Shaping a Resilient and Sustainable Energy
- Future
- Glossary
- *Agreement and Treatise*
- Reference
- About the Author

DEDICATION

To the tireless visionaries who dare to reimagine a stable and sustainable world,

To the countless individuals working behind the scenes to illuminate our homes, fuel our progress, and secure our future, even in the face of uncertainty,

And to my beloved family, whose unwavering support, patience, and encouragement have been the energy that sustained me throughout this journey—

This book is for you.
May it serve as a beacon of hope and a call to action in navigating the path from crisis to stability.

ACKNOWLEDGEMENT

Writing From Crisis to Stability: Understanding Energy Security in a Volatile World has been an extraordinary journey, one that would not have been possible without the support, guidance, and inspiration of many remarkable individuals.

First and foremost, I am deeply grateful to God who is my source of inspiration for the wisdom and grace and constant source of strength throughout this project.

To my family, your unwavering support and patience have been my anchor. To my spouse, Ruth, your encouragement and belief in me have fueled my resolve, even on the most challenging days. To my children, your curiosity about the world reminds me of why understanding energy security matters for the future.

To my colleagues and mentors, thank you for sharing your insights and expertise, which shaped the foundation of this book. Your critical feedback, stimulating discussions, and shared passion for global stability have enriched this work beyond measure.

I am profoundly grateful to the researchers, policymakers, and energy experts whose work I have drawn upon. Your commitment to navigating the complexities of energy security in an ever-changing world has been a source of inspiration.

A heartfelt thanks to my editor and publishing team, whose meticulous attention to detail and belief in the vision of this book helped transform a collection of ideas into a cohesive narrative.

Finally, to my readers, this book is for you. In a world where energy security shapes economies, livelihoods, and geopolitics, your engagement with these issues is crucial. I hope this work sparks meaningful conversations and inspires actionable solutions for a more stable and sustainable future.

With gratitude,
Olugbenga Onungbogbo

PREFACE

Energy security is a critical issue that touches every corner of our lives and it is not just a concern for policymakers or industry experts, from the lights in our homes to the fuel that powers our economies, energy is the lifeblood of modern civilization. Yet as we stand at the crossroads of the 21st century, the stability of this lifeblood is under serious threat.

It was Winston Churchill who famously said, "Safety and certainty in oil lie in variety and variety alone." This statement made during the early days of the oil industry resonates even more today. In a world where geopolitical tensions, climate change and technological vulnerabilities intersect with our growing energy demands, the need for a diverse and resilient energy supply has never been more urgent.

As I embarked on writing this book, my aim was to delve deep into the complexities of energy security and to shed light on the multifaceted challenges we face. The energy crises of the past have shown us how vulnerable our global systems can be, but they have also taught us valuable lessons about resilience, innovation, and the power of cooperation. "From Crisis to Stability" seeks to explore these lessons and provide a roadmap for securing our energy future.

As the world pivots toward a more sustainable energy landscape, we must remember the words of Albert Einstein: "We cannot solve our problems with the same thinking we used when we created them." The transition to renewable energy, the protection of critical infrastructure and the pursuit of global cooperation require new, bold strategies that go beyond traditional approaches.

This book is not just a scholarly exploration; it is a call to action, whether you are a policymaker, an industry leader or a concerned global citizen, the time to act is now. Our collective future depends on our ability to ensure that energy is secure, sustainable, and accessible for all. Together we can move from crisis to stability, forging a path toward a brighter more resilient future.

The power to secure our energy future lies in our hands. Let's make sure we don't drop the ball.

Olugbenga Onungbogbo
Abuja, Nigeria

CHAPTER ONE

THE HISTORICAL EVOLUTION OF GLOBAL ENERGY SYSTEM

"The Stone Age didn't end because we ran out of stones."

SHEIKH AHMED ZAKI YAMAN

Indeed, civilization has been driven by energy-from the earliest of times to the modern-day industrial age-in the unending pursuit for energy. It shaped societies, fueled economic growth, and defined the geopolitical landscape of territories. Understanding the history of global energy systems provides a necessary foundation for tackling the energy security problems of today and tomorrow. This chapter traces the history of the development of energy systems from their beginnings in primitive forms of power to the complex, interconnected worldwide energy system that supports our present society.

The Emergence of Energy Utilization

Human beings have always relied on energy as a basic survival need. Early humans living during prehistoric times used the very simplest form of energy, fire. Fire was a revolutionary discovery that allowed for the cooking of food, warding off predators, and sources of warmth. This very rudimentary, deep-reaching energy source permitted early societies to migrate into colder climates and establish communities in previously uninhabitable regions. As these early communities grew, so did their energy needs; the first significant shift in energy use occurred with the advent of agriculture around 10,000 BCE. Animal domestication and tool use greatly expanded the yield on food production, allowing for population growth and more permanent settlements. These agricultural societies still heavily depended on biomass-wood, crop residues, and animal dung-as their mainstays of energy. Biomass was burned for cooking, heating, and later for limited industrial purposes, such as pottery and metallurgy.

The Transition to Fossil Fuels: Coal and the Industrial Revolution

All through many millennia, human society was heavily reliant on muscular energy—both human and animal—work, and biomass for heat and cooking; it drastically turned an about-face with the coming of the Industrial Revolution back in the 18th century. The development of steam engines with coal as a propellant ushered in a new era in energy use.

While coal-remains of ancient plants and animals formed over millions of years-became the cornerstone of the Industrial Revolution, the first practical steam engine was developed by Thomas Newcomen in 1712 to pump water out of coal mines. Improvements by James Watt in the late 18th century made steam engines much more efficient and flexible, their application became absolutely general, these were installed in factories, ships and trains. Economies and societies were transformed by steam-powered engines.

The use of coal as a source of energy was more than an economic transition; it had enormous social and environmental consequences, too. This forced people to move to cities in search of jobs in the coal-powered factories, and thus, urbanization gained momentum. The cities were, unfortunately, often plagued by acute pollution since the burning of coal produced a great deal of smoke and soot which contaminated the atmosphere. These environmental effects stemming from coal use served as an early warning for what was later to define energy security in the modern world.

The Emergence of Oil: History of the 20th Century

The 19th century was the century of coal; it was followed, almost immediately, in the 20th century by the rise of oil as the dominant source of energy. Starting with the first well drilled by Edwin Drake in Pennsylvania in 1859, the oil industry was born. Oil's early uses were limited to kerosene lamps, but it grew in significance with the internal combustion engine and automobile development.

In the early 20th century, the oil industry rapidly expanded. With the invention and then mass production of automobiles by firms like Ford, oil became the preferable fuel for transportation. It is this mobility afforded by oil-powered transportation that revolutionized societies, enabled suburbanization, and totally reshaped urban landscapes. Most important, the rise of aviation entrenches oil's predominance, since airplanes became a vital means both for civil aviation and military purposes.

The strategic relevance of oil was acutely evident in World War I and World War II. Control of oil resources played a critical part in military strategy, and the desire for secure supplies of oil underlined the geopolitical action of states. The Middle East, with its tremendous oil supplies, became an essential zone of attention for global politics, a position it still occupies today.

Together with the predominance of oil, new challenges also arose during the 20th century. The environmental impact of extraction and consumption of oil-oil spills, air pollution, and greenhouse gas emissions-became a source of concern. Furthermore, the supply of oil was highly concentrated in a few regions, mainly the Middle East, which posed a threat to the world's energy security. The oil shocks of the 1970s, driven by geopolitical tension and the oil embargo slapped by OPEC, brought into relief a worldwide dependence on oil and accompanying risks.

The Nuclear Age: Promise and Peril

In the mid-20th century, as the world struggled with its dependence on oil, a new form of energy came into view: nuclear power. The discovery of nuclear fission in the 1930s, and subsequently the development of nuclear reactors, seemed to offer a virtually unlimited source of energy. Nuclear power plants, which first came online in the 1950s, were capable of generating enormous amounts of electricity with very little environmental impact-at least, in terms of carbon emissions.

Initial enthusiasm for nuclear power was partly based on the promise of energy security since the fuel could be produced in many countries. This reduces or avoids dependence on foreign energy sources, unlike finite supplies of fossil fuels concentrated in specific regions. Additionally, it was meant to meet the increasing demand for energy without adding to air pollution and greenhouse gas emissions.

However, the promise of nuclear energy became soured by the risks it entailed. The disasters at Three Mile Island in 1979, Chernobyl in 1986, and Fukushima in 2011 underlined dangers from nuclear power. These events have enhanced public fears about the safety of nuclear energy, leading to a slowdown in constructing new nuclear plants and decommissioning in some countries.

The question of nuclear energy is similarly controversial. Low-carbon, though it may be, and an alternative to fossil fuels, the risk of accidents, radioactive-waste management, and nuclear proliferation keep substantial barriers in place. In an era when energy security has to be weighed against environmental sustainability, the contribution that nuclear power will make becomes a vital question. Birth of Renewable Energy: Solar, Wind, and Beyond

Renewable energy has emerged as one of the most important parts of the global energy mix since the late 20th and early 21st century. Gained consciousness on climate change and other environmental impacts caused by the use of fossil fuels increased the attraction toward energy sources that were clean, abundant, and sustainable.

First solar, then wind power have been the leading agents in the renewable energy revolution. The development of photovoltaic cells in the 1950s laid the foundation for modern solar power technology. From early, highly specialized use-like powering satellites-solar power has dramatically scaled. With improved technology and decreasing costs, solar power today is becoming a progressively competitive option to sources of traditional energy.

Wind power has also shown very keen growth, with wind turbines now a common enough sight in large parts of the world, converting the kinetic energy of the wind into electricity. The scalability of wind power-from small, individual turbines to large offshore wind farms-has made it popular to diversify energy portfolios and help reduce carbon emissions.

In addition to solar and wind, other renewable resources such as hydroelectric, geothermal, and biomass supplement this energy deficiency in many parts of the world. Hydroelectricity has been exploited for over a hundred years and is still a significant source of electricity in many regions worldwide. Geothermal energy is another rich resource derived from the thermal energy of the Earth; these sources are readily found in select regions where the geology permits. Biomass-energy produced through organic materials-therefore represents a renewable resource compared to fossil fuels but also, because of the implications for land use and carbon emissions, presents challenges. Renewables are restructuring the global energy system. Driven by the twin goals of reducing greenhouse gas emissions and enhancing energy security, countries around the world are investing in renewable infrastructure. There are, however, also challenges to transition toward a future of renewable energy, and several issues would have to be dealt with in the process: integration of renewable energies into power grids, management of intermittency, and proper availability of critical materials used in renewable technologies.

The Energy Landscape Today: Complex and Interconnected

Today, the energy landscape is more intertwined and complex than it has ever been in the history of humanity because the energy system that powered our world was a mosaic of traditional fossil fuels, nuclear power, and renewable sources of energy, and this diversity brings into our world both opportunities and challenges.

While on the one hand, diversification with a variety of energy sources improves energy security by decreasing dependence on any one single source. On the other side, global energy markets are so interlinked that disruptions in one part of the world are felt very far and wide into another part of the world. This can be understood by the fact that manifold risks are associated with volatile oil prices, geopolitical tensions affecting energy supply chains, and energy infrastructure vulnerable to cyberattacks, among other factors. Besides, a host of external factors singly and jointly shapes the energy landscape, which includes but is not limited to economic trends, technological innovation, and environmental policy. The low-carbon economy, which is compelled by the urge to mitigate climate change, will basically alter the production and consumption of energy. Meanwhile, in other arenas, an evolving technology arena is opening new opportunities toward improved security and sustainability of energy, from smart grids to energy storage. Conclusion: The Path Forward

History is in a nutshell the unwavering resolve of humanity for progress, with the transformational capability of energy to shape our world-from that very first controlled use of fire to today's complex global networks of energy. It has been a journey of continuous adaptation, innovation, and discovery. Each transition-from biomass to coal, from coal to oil, and toward today's more diversified and sustainable mix of energy-has been propelled by imperatives of meeting rising demands, surmounting setbacks, and making the most of new opportunities. It has bequeathed on us not insignificant challenges-from environmental degradation and geopolitical conflict to the ever-imminent threat of energy insecurity. These are increasingly complex, interconnected problems that are pressing in urgency as we look to the future. The global energy landscape stands at a critical juncture-what happens today will determine whether we can secure a stable and sustainable energy future.

Our way forward needs to learn from history but also embrace innovation. We need to invest in diverse and robust energy systems that would adapt to the shifting circumstances of life and be resistant to various shocks. This would mean the acceleration of renewable energy transition, energy efficiency, and accompanying technologies like energy storage and smart grids. Meanwhile, we also have to address the vulnerabilities of our existing infrastructure in order not to be susceptible to various physical and cyber threats. Besides, navigating the complex theme of global energy security, international cooperation is going to be paramount. The problems transcend borders, and so do their solutions. Only collective actions will lead informed by commonly held values on sustainability, equity, and resilience toward a certain and sustainable energy future.

As we forge ahead, let us remember energy security does not stop at the question of a reliable supply of energy. Rather, it goes toward creating a system that would foster human development, protect the environment, and promote peace and stability. The road to such a future will by no means be an easy one, but it is a road we must travel with determination, innovation, and commitment to a better world for all.

CHAPTER TWO

THE DIMENSIONS OF ENERGY SECURITY

"Energy is the golden thread that connects economic growth, increased social equity and an environment that allows the world to thrive"

Ban Ki-moon

Energy security is a multidimensional concept that goes beyond the simple availability of supplies of energy. It is constituted of several dimensions, all important in their own respect in ensuring that societies have access to energy in a reliable, affordable, and sustainable manner. In this chapter, the dimensions of energy security shall be taken into consideration in discussion, considering four key aspects of availability, affordability, accessibility, and sustainability. These dimensions, if understood, would provide a better grasp of the complexity of energy security and the challenges that must be addressed to achieve it. Availability: Ensuring a Steady Supply of Energy
Probably the most basic dimension of energy security is availability. It concerns the physical availability of energy resources-fossil fuels, renewable energy, or nuclear power-and the infrastructure to produce, transport, and distribute energy. Without a regular and dependable supply of energy, the functioning of economies and societies, not to say national security, can be severely compromised.

Several factors bear upon the availability of energy, and these include natural resource endowments, technological capabilities, and geopolitical conditions.

Countries that have oil, natural gas, or coal in abundance enjoy a strategic advantage in the context of energy security. Put differently, even those countries that possess enormous resources may also face problems when they cannot support effective infrastructural needs for extraction, processing, and transportation of these resources to where they shall be put into use. For example, Russia represents one of the largest reserves of natural gas and oil in the world, which positions it as a key participant in the global energy market. On the one hand, the country has rich energy resources; on the other hand, its energy security relies deeply on a broad network of pipelines and export facilities throughout the country. Natural disasters, technical failures, or political conflicts stoppages could have great implications for the availability of energy within Russia and countries dependent upon its energy exports.

On a global scale, energy supply and demand dynamics come into play. The sudden boom in the global demand for energy caused by population increase, urbanization and economic development makes sure that the supply of energy is unquestionably difficult to maintain in regions of the world where energy resources are scarce or difficult to access, such as in the case of remote areas or deep-sea oil reserves.

Though technological advances are also playing a critical role in improving energy availability, new exploration and extraction technologies, such as fracking and deep-sea drilling, have made resources available that were previously inaccessible for exploitation. Likewise, renewable energy technologies that are alternative to fossil fuels, like solar panel and wind turbine, have increased

clean energy production. Therefore, these alternative fuel technologies have their own set of challenges to meet. Today, environmental impact, high cost, and huge infrastructural investment remain some of the most significant barriers to solar panel and wind turbine development.
Affordability: The Economic Dimension of Energy Security

Of course, the availability of energy is a critical factor, but in itself does not guarantee energy security. Energy has to be available at an affordable price to consumers, businesses, and governments alike. To this end, it indicates that people and societies should have access to energy sources without improper economic burdens.

Energy affordability is also closely related with the dynamics of the energy market concerning supply and demand, cost of production and distribution, and policy and regulations governing the energy market. Large changes in the price of energy can have significant impacts both economically and socially. For example, steep increases in oil prices-as during the two oil crises of the 1970s-have led to recession, inflation, and social unrest.

Energy poverty is one of the most important issues in many countries due to a high energy price increase, which would put the low-income household into a dilemma between heating their homes and other vital expenses like food, healthcare, and education.

This phenomenon, generally labeled "energy poverty," affects several developing countries where access to modern energy services is usually minimal. However, it is Governments that play a crucial role in ensuring energy affordability through the adoption of policies and regulations influencing energy prices. Subsidies for fossil fuels, renewable energy incentives, and price controls are some of the tools governments use in trying to maintain stable and reasonable energy prices. But these measures frequently have side effects, such as distorting market signals, encouraging profligate consumption, and burdening public finances. For example, many oil-exporting countries, including Saudi Arabia and Venezuela, have long subsidized their domestic fuel prices so that energy can remain inexpensive for their citizens. While these subsidies have helped promote social stability, they have simultaneously helped cause overconsumption, reduced incentives to promote energy efficiency, and strained government budgets-particularly when world oil prices fall. Transition to renewable energy has also brought along issues of affordability, whereby while the costs of renewable energy technologies, such as solar and wind power, have drastically gone down in recent years, its upfront investment requirement remains high.

Moreover, the intermittency of renewable energy sources-solar and wind-increases the energy storage and integration costs into the grid. Ensuring this transition to a low-carbon energy system is available and fair will be one of the biggest challenges for the coming decades. Accessibility: Reaching Every Corner of Society

Energy accessibility refers to the availability of enough energy at an affordable price for different populations, communities, and regions. This dimension of energy security is directly related to equity issues, social justice, and development. The International Energy Agency estimates that access to modern energy services is limited or absent, especially in developing parts of the world, resulting in large differences in quality of life and economic opportunities. An estimated nearly 770 million people around the world still lack access to electricity, and many more rely on traditional biomass-such as wood, charcoal, and dung-for cooking and heating. This lack of modern energy services has deep implications for health, education, gender equality, and economic development. For example, in sub-Saharan Africa, where energy access remains among the poorest regions of the world, most households use biomass for cooking, leading to indoor air pollution from the combustion of plants and animal dung, which causes respiratory illnesses and early deaths. Women and children generally bear the burden of gathering firewood-a very time-consuming and exhausting job that it leaves them little or no chance to pursue education and economic participation. In this direction, access to energy has become one of the priorities in international development. The United Nations Sustainable Development Goal 7 ascertains that all people have access to affordable, reliable, sustainable, and modern energy until 2030. In fact, meeting this goal will take a great investment in energy infrastructure, especially in rural and far-flung areas of a country, and innovation in off-grid and decentralized systems of energy.

Decentralized energy systems, such as solar home systems and mini-grids, present a very good prospect for increasing accessibility of energy supplies in areas where central grid extension is either not feasible or not cost-effective. These can ensure adequate, reasonably priced energy supply to communities currently unserved, thus decreasing the consumption of fossil fuels and lowering GHG emissions. However, increasing accessibility of energy is not limited to infrastructural expansion alone but also to the large-scale social, economic, and cultural domains inhibiting proper energy access.

The fact is, for example, that even in areas where there is infrastructure supplying energy, many poor households cannot afford it. In addition, access to energy can be influenced by various cultural norms and gender roles that dictate who has access to energy and for what use. These are some of the outstanding barriers to overcome, and doing so will require an integrated approach that considers various needs and perspectives of all members of society. Sustainability: Balancing Energy Security with Environmental Stewardship The sustainability dimension in energy security increasingly gains recognition within the context of global climate change and environmental degradation.

The definition of sustainability here is the degree to which present needs in energy can be met without compromising the ability of future generations to meet their own needs.

It involves environmental impacts of energy production and consumption, including greenhouse gas emissions, air and water pollution, and resource depletion. Active Fossil Fuels - that is, Coal, Oil, and Natural Gas-are the dominant sources of energy over the last century. However, it is very well-known that their use obviously comes with huge environmental burdens: burning fossil fuels represents the largest source of carbon dioxide (CO_2) emissions, mainly responsible for global warming. Besides, extracting and processing fossil fuels can result in habitat destruction, water contamination, and other environmental damages. It is by now widely recognized that the need exists to move toward a more sustainable energy system, but such a transition creates some very substantial barriers. The alternative sources of renewable energy-solar wind, and hydropower-are cleaner options compared to fossil fuel sources, but all have varying environmental impacts. Very large-scale hydropower projects can disrupt ecosystems and displace communities, while the production of solar panels and wind turbines requires raw material extraction and waste generation. Apart from reducing the environmental impact brought about by the generation of energy, sustainability also means enhancing energy efficiency. Energy efficiency refers to an act of using less energy in performing a similar service or producing the same output. Consequently, improving energy efficiency reduces energy consumption that, in turn, makes less GHG emission and diminishes energy resources demand in general.

The action for energy sustainability can be enacted in governments, businesses, and individuals through various ways: setting policies and providing incentives by policymakers to stimulate renewable energy and the application of energy-efficient technologies. Businesses may invest in clean energy innovations and implement sustainable practices in their operations. Besides, it can be enacted through reduction of energy use and carbon footprint by consumers' choice to use energy-efficient appliances, drive fuel-efficient vehicles, and choose renewable energy providers.

This will not only have implications for the transition to a sustainable energy system but also for the energy security aspect. On one hand, a reduced dependence on the sources will enhance energy security by diversification of the energy mix and reduction of exposure to volatile global energy markets. In their stead, transition towards renewable energy brings in new challenges: energy storage and grid integration, availability of critical minerals, and land-use conflicts.

The Dimensions of Energy Security Interconnectedness

These four dimensions of energy security-availability, affordability, accessibility, and sustainability-are by no means mutually exclusive. Each impacts and informs the others in a multi-dimensional and often contradictory manner. Addressing one dimension often has consequences for the others, creating a set of challenges and opportunities in the quest for holistic energy security. For instance, increasing the supply of fossil fuels to improve energy availability can also undermine sustainability by increasing greenhouse gas emissions and other environmental degradation, perhaps reducing long-term energy security as climate change impacts begin to curtail resource availability and destroy infrastructure. Similarly, measures for pursuing 'affordable' energy may at times work against sustainability.

Giving subsidy support to fossil fuels for affordable prices will only result in higher consumption and delayed transition towards cleaner sources of energy. On the other hand, investment in renewable energy is likely to increase costs in the short term due to high upfront expenses; however, in the long term, it allows for a circle of affordability through lower operational expenses and independence from volatile fossil fuel markets. Not only is access closely intertwined with other dimensions, but an increase in access to energy in underserved regions could have spillover effects on economic development and social equity. That all will have this energy in a sustainable manner does, of course, demand planned consideration to avoid degradation of the environment and also ensures energy remains affordable to all. Interconnectedness among these dimensions underlines the complexity of achieving true energy security. These will have to be taken into consideration at the design of strategies by policymakers, businesses, and communities, realizing that gains in one area should not be at the expense of others but are a balanced manner, one seeking to harmonize availability, affordability, access, and sustainability, which is what will deliver an energy system that will not only be secure but fair and resilient for the next generation.

CHAPTER THREE

THE GEOPOLITICS OF ENERGY SECURITY

"Whoever controls the energy can control whole continents"
Henry A. Kissinger

Energy security is thus both a national issue and an international concern-a concern impregnated with the same complexities and potential conflicts evident in international relations. In this respect, geopolitics refers to the strategic imperatives, policies, and relationships that exist among nations in their competing efforts to grapple with fulfilling their energy needs through rivalries, cooperation, and conflict.

This chapter will, therefore, consider the geopolitical perspective of energy security and explain how access to energy resources, management of supply routes, and dynamics of the global energy market could influence international relations and affect global stability.

Energy Resources and Geopolitical Authority

In historical terms, energy resources have often been a fulcrum of geopolitical power. Countries that are well endowed with such resources, especially oil and natural gas, have often demonstrated numerous ways of making their presence felt in the international arena. Such endowments may act as an instrument of diplomacy, economic leverage, and at times, coercion. Control over energy resources and their distribution enables certain countries to shape world energy markets, influence other states' policies, and protect strategic interests.

The Middle East is generally typified by large reserves of oil and natural gas, a perfect example of how energy resources can confer geopolitical influence. For decades, countries within the region have, of course, played a role in global energy markets: Saudi Arabia, Iran, Iraq, Kuwait, among others, has served as a key supplier of oil to the rest of the world. This is where a consortium of oil-exporting countries under the name Organization of the Petroleum Exporting Countries has been able to influence world prices of oil by coordinating production rates among member states.

Of course, such geopolitical leverage coming with energy resources only brings a load of challenges. Historically, the Middle East has been at the center of much geopolitical tension, where most of it relates to the control over resources of energy. Armed conflict, political instability, and regional rivalries have frequently disrupted oil production and its exports, thereby creating volatility in the global energy markets. For example, the 1973 oil embargo saw OPEC member states use oil as a geopolitical tool; this led to an energy crisis around the world, creating widespread economic and political effects.

Other regions outside the Middle East have also used energy supplies as part of geopolitics. For example, Russia has used its rich resources in natural gas to increase its influence among its neighbors, most of them in Eastern Europe. Countries like Ukraine, which is highly dependent on Russian gas, have appeared to become more politically vulnerable to pressure from Moscow, as illustrated by conflicts over gas supply in the mid-2000s and in the ongoing war between Russia and Ukraine. These conflicts have affected not only Ukraine but also had larger spillovers across Europe, underlining broader geopolitical implications of energy dependence.

The rising energy resource needs of China have dramatically changed the nature of global geopolitics. As the world's most dominant energy consumer, China has sought to secure access to energy resources around the world, especially from Africa, the Middle East, and Central Asia. This has been accompanied by a strong increase in Chinese investment in energy infrastructure and the extraction of resources within these regions, many times tied to growing political and economic influence. The "Belt and Road Initiative" by China, a hugely ambitious infrastructure and investment plan, has significant energy components to underpin the country's strategic priority of securing energy supplies for its economy.

Logistical Pathways and Critical Strategic Constraints

If energy resources are considered critical, then the routes through which they are transported assume equal importance in the geopolitics of energy security. The strategic chokepoints-narrow passages that include straits, canals, and pipelines-form a critical link in the global supply chain of energy. Hence, control over such chokepoints gives any country significant geopolitical leverage, while vulnerability to disruption creates a serious security concern for energy-importing countries.

The Strait of Hormuz has become a crucial strategic chokepoint at the global level, forming a slim sea connection between the Persian Gulf and the Arabian Sea. It is assumed that around 20% of the world's oil passes through this vital waterway, and therefore, it is an essential route from the point of view of global energy supply. Its strategic importance means that any disruption in shipping through this area-military conflict, terrorism, or whatever-can have dire consequences for the global energy markets and overall international stability.

The Strait of Hormuz attained strategic significance and became a focal point of geopolitical rivalry. Iran, a state bordering the strait, has from time to time threatened to block it in response to sanctions or military pressures applied by Washington and its allies. Such threats underscore the vulnerability of international energy supplies to possible geopolitical discord and the need to protect vital supply routes.

Another essential chokepoint is the Suez Canal, providing a connection between the Mediterranean Sea and the Red Sea and enabling a non-stop route for oil and LNG shipments from the Middle East to Europe and North America. The Suez Canal forms one of the critical elements in the global energy supply chain, and since this canal is required for shipping significant volumes, any disruption-whether due to conflict, accidents, or other disorders-has the potential to cause significant hold-ups and increased transport costs.

Its pathways in turn would greatly affect the geopolitics of energy security. Pipelines, highly efficient ways to transport oil and gas over land, remain perennially prone to acts of sabotage, political volatility, and geopolitical conflict. A specific example of this includes the Nord Stream pipelines, through which natural gas flows from Russia to Germany beneath the Baltic Sea. This has become the focal point of geopolitical strife between Russia, Europe, and the United States. Fears over Europe's reliance on Russian gas have prompted debate on the security implications of these pipelines and efforts to diversify supplies.

Pipeline routings have become a main determinant of geopolitical competition in regions such as Central Asia and the Caucasus. Pipeline infrastructure has been crucial to the exportation of natural resources from countries like Azerbaijan, Kazakhstan, and Turkmenistan to foreign markets. The directions taken by these pipelines-to pass either through Russia, Turkey, or others-are usually delineated by geopolitical forces of alliances, conflicts, and economic interests.

The Global Energy Market and Economic Interdependence

The international energy market is a significant sector of geopolitics that concerns energy security. Energy is a traded good in the world market, whereby interdependence of countries opens up avenues for cooperation, and at the same time, it opens up possibilities for conflict. In the global energy market, dynamics feature in relation to the following aspects: supply and demand, price fluctuations, changes in technology, and geopolitical events. Another critical determinant of the global energy market is the impact brought about by the price of oil, since this is the most widely traded energy commodity. Therefore, this would have a much greater consequence on the world economy. These changes can be brought about by numerous variables, which also include changes in volumes of production, geopolitical events, natural disasters, and shifting demand. The steep decline in prices of oil in 2014, for example-driven by a mix of increased U.S. shale oil production and a decision by OPEC not to cut output-had wide-ranging economic consequences for oil-exporting countries like Russia, Venezuela, and Nigeria.

The international energy market is a good example of the increasing interdependence of nations in relation to energy production, consumption, and trade. Interdependence in this way may allow opportunities for cooperation in efforts by nations together to ensure stability in energy supplies and markets. A typical example is the creation of the International Energy Agency-IEA, which was a response to the oil crisis of 1973 for the purpose of developing energy security and cooperation among its member states. The IEA co-ordinates emergency response measures such as strategic oil reserves to mitigate the impact of supply disruptions.

After all, the global energy market interdependence brings with it some very vexing problems, as countries reliant upon energy imports bear a host of heightened risks when it comes to disruptions to supply, price volatility, and even geopolitical instability. For instance, Japan, being extremely import-dependent to meet all of its virtually diverse energy needs, is highly sensitive to changes in the world energy markets and to the security of its supply routes. Similarly, European countries dependent on Russian gas stand to suffer from geopolitical disputes that could disrupt their supplies.

At the same time, a more diversified and sustainable energy portfolio is transforming the global energy marketplace. Expanding renewable energy sources, new technologies such as electric vehicles and energy storage, and the greater emphasis on efficiency are reconstituting the business forces of energy. These changes are providing new opportunities for countries to further decrease their dependence on fossil fuels, strengthen their energy security, and minimize environmental impacts associated with energy supply and use. For instance, the growth in renewable energy sources would decrease the geopolitical power of traditional energy-exporting states by reducing global demand for oil and gas. Similarly, value creation within renewable energy technologies, such as solar panels and wind turbines, is concentrated mainly in a few countries, giving rise to novel dependences and geopolitical consequences. For instance, China is a leading producer of solar panels and rare earth minerals, key materials for many renewable energy technologies. That places China in a strategic position within the global transition toward clean energy, at the same time as it heightens concerns about the further consolidation of supply chains.

Energy Security and Global Governance

Bearing in mind that energy security has global dimensions, constructive international cooperation and governance to mitigate problems and risks associated with the geopolitical dynamics of energy is a necessity nowadays. The international governance structures that encompass international organizations, treaties, and multilateral agreements have a great potential for promoting principles of energy security, stability, and sustainability. A major agent involved in energy security matters is the International Energy Agency that brings together its member states for the unification of energy policies and sharing of best practices in response to energy crises. The IEA mandate has grown steadily to cover, inter alia, energy security attention to energy efficiency, renewable sources of energy, and avenues towards the mitigation of climate change. The ECT is another significant international tool which was created with the purpose of encouraging energy cooperation, investment protection, and energy transit across borders. Because the ECT sets up a legal framework for dispute settlement between countries and investors in the energy sector, it contributes to stability and predictability in the global energy market.

However, international governance on energy security is often problematic due to countries having different interests and focuses. While some countries are focused on energy security and economic growth, other countries maintain focus on environmental sustainability and reduction of climate change. Such differences may cause conflicts in international negotiations, as has been seen in the negotiations on climate change agreements.

CHAPTER FOUR

THE ROLE OF TECHNOLOGY IN ENHANCING ENERGY SECURITY

"The future of energy is not in oil or coal, but in wind, solar, and other renewable resources"
                                                          Gina McCarthy

Energy security is surely one of the critical concerns of the 21st century, with the increase in energy demand and geopolitical tensions, while the transition toward more sustainable sources of energy is becoming imperative. Therefore, technology can thus play an ever-more central role as nations seek to guarantee themselves stable, affordable, and clean energy supplies. New technologies are transforming not just how energy is produced, stored, and distributed, but also how challenges to energy security are addressed. The chapter will review the main technological advances that mark the transformation of the energy world, with a special view to renewable energies, energy storage, smart grids, and capture, utilization, and storage of carbon dioxide (CCUS).

Renewable Energies Technologies

The shift to renewable energy thus forms the core of the world's strategy for improving its energy security. Unlike fossil fuel, which is finite and concentrated in specific regions, renewable sources of energy, such as solar, wind, and hydropower, are abundant, dispersed, and inherently cleaner. Technological innovations have dramatically reduced costs and improved the efficiency of the technologies, making them increasingly competitive with traditional sources of energy.

Solar Power: Solar energy has grown remarkably in recent times due to technological development in the area of photo-voltaic cells. These cells directly convert sunlight into electrical energy, and their efficiency is now much better compared to earlier times, while costs of production have decreased. Large-scale solar farms and rooftop solar panels for residential and commercial purposes have been deployed rather rapidly. The leading positions in terms of solar power capacity attained by countries such as China, the United States, and India already contribute much to energy security by saving imports of fossil fuels. Besides, the decentralized nature of solar energy permits localized energy production in a manner that reduces vulnerability to disruptions in either centralized power generation or supply chains.

Wind Energy: Wind energy is one of the critical parts of the renewable energy mix that is currently available and deployable as a scalable, affordable solution to energy security. The evolution of continuous technological improvements in turbine design, materials, and deployment strategies has given rise to ever-higher capacities and further efficiencies. Offshore wind farms, now gaining increasing popularity, especially in Europe and Asia, offer better efficiency compared to onshore due to stronger and more consistent winds. The integration of wind energy, therefore, brings diversity into the energy sector and cuts overdependence on a single source of energy. Besides, wind farms are normally modular; hence, their capacity can be built in increments to attain flexibility in scaling up energy production in response to demand.

Hydropower and Geothermal Energy: Hydropower remains the largest source of renewable electricity globally, providing reliable and consistent energy. Advancement of small-scale hydropower technologies makes this technology viable for more extensive applications, mainly in those areas which have become distant from the center where access to the grid is very minimal. Another promising form of renewable technologies is geothermal energy: heat emanating from inside the earth harnessed as. While at present its deployment is restricted to regions of the world with favorable geological conditions, innovations in enhanced geothermal systems are increasing the places where this technology can be deployed; hydropower and geothermal power have a further advantage over these other two, since they can supply a firm baseload of electricity, complementing the solar and wind energies, which supply power only intermittently.

New Renewable Technologies

Future Storage Technologies: Besides already well-established technologies such as batteries and pumped hydro, a number of newer energy storage technologies are highly promising in ensuring energy security. Liquid electrolytes-based flow batteries hold considerable promise on longer-duration and scalability compared to lithium-ion batteries; hence, they fit in perfectly with the existing grid-scale applications. Other technologies under consideration include compressed air energy storage and flywheel energy storage, variably large-scale and long-duration storage capability. Thermal energy storage, which stores energy in heat form, is gaining attention due to applications in concentrated solar power plants and industrial processes. These are critical technologies that enable the management of renewable energy variability with an eye toward stability and reliability of energy supply.

These include ocean energy, advanced bioenergy, and hydrogen-technologies that afford promising opportunities for further environmental sustainability. Ocean energy is harnessed from ocean currents and waves through tidal and wave power, making it a consistent and renewable energy source. Advanced bioenergy, derived from sustainable biomass sources, offers low-carbon alternatives to fossil fuels. Hydrogen-especially when produced from renewable sources, better known as green hydrogen-is a clean energy carrier with a number of applications, including producing power and providing fuel for transport.

Energy Storage Technologies: Due to the intermittent nature, integration of renewable energy to the grid poses a major challenge since it offers higher or lower renewable generation. Energy storage in the energy sector is emerging with advancement for the effective balancing of supply and demand, along with increasing the reliability of sources of solar and wind power. Energy storage technologies are therefore crucial in addressing the challenge in that they are capable of storing excessive energy during high production and release it at times when demand for renewable energy is low. Battery Storage: The lithium-ion batteries are the most deployed energy storage technology today, which offer high energy density, efficiency, and scalability. The reduction in the cost of batteries driven by both technology and economies of scale has finally made the technology relevant for battery storage at both the grid-scale and distributed systems. The battery storage systems provide essential grid services like frequency regulation, voltage support, and peak shaving that are necessary to maintain the stability of the grid. Batteries also make it possible to include more renewable energy because excess generation is reserved and used at a time when generation of this form is not possible, thereby cutting the use of fossil fuel-based backup power. Pumped Hydro Storage Pumped hydro storage currently represents the most mature form of large-scale energy storage and constitutes the majority of global energy storage capacity. PHS works by using excess electricity to pump water from a lower reservoir to an upper reservoir. In times of peak demand for electricity, this stored water is released to flow back down through turbines, creating the generation of electricity. This results in considerable grid balancing capabilities, as PHS is suitable to complement variable renewable energy sources. In mutual limitation, geographical and environmental constrains set limits to its terrain and deployment of PHS-it requires suitable ones which can have significant ecological impacts.

Hydrogen as an Energy Carrier: The role of hydrogen has become an important factor for the future energy system, with broad perspectives on energy storage, transportation, and decarbonization. Green hydrogen, if derived from renewable sources, can store renewable energy for a longer period and thus facilitate supplies buffering seasonal demand and supply fluctuations. Hydrogen can also be utilized in delivering power, transport, and industrial processes, which would lessen the dependence on fossil fuel consumption and increase energy security. In order to fully realize the potential of hydrogen within a low-carbon energy system, efficient development in areas such as electrolyzers, fuel cells, and hydrogen storage systems will be required.

Smart Grids and Digitalization

Other key factors contributing to energy security improvement involve upgrading the electricity grid through smart enablers and digitalization. Traditional power grids were designed for the centralized one-way flow of electricity from large power plants down to the consumers. However, increased deployment of DERs like rooftop solar panels, electric vehicles, and energy storage systems has created a dire need to shift toward more flexible and dynamic grid architectures.

Smart grids deploy advanced communication, automation, and data analytics to optimize the operation of the electricity system, improve reliability, and integrate renewable energy sources more effectively. These will be grids with two-way flow of power-that is, while consumers may draw electricity from the grid, they can also supply electricity back into the grid. This also makes energy production more decentralized, improving energy security because the chances of a network failure decrease, and energy can be produced locally where it is consumed. Smart grids enable integration of DERs, a strong enabler towards a resilient and adaptive energy system.

Demand-Side Management: Smart grids further enhance energy security by making demand-side management possible, where consumption patterns change in accordance with supply conditions. DSM allows utilities to better balance supply and demand, decreasing reliance on expensive and polluting peaking power plants. Several enabling technologies for consumer participation in DSM programs are smart meters, connected appliances, and home energy management systems that contribute to grid stability and efficiency. By shifting electricity use to off-peak times or reducing consumption during periods of high demand, DSM helps prevent grid overloads and reduces the likelihood of blackouts.

Resilience and Cybersecurity: Digitalization of energy makes the development of resilient and cyber-secure smart grids of paramount importance. In addition, cyberattacks on critical infrastructures such as the electric grid will have dramatic consequences for energy security. Advanced cybersecurity measures, including encryption, intrusion detection, and AI-based threat monitoring, have made the smart grid resilient in case of cyber threats. Smart grids enhance physical resilience through faster outage detection and response resulting from better technologies and designs, which reduce the impacts of natural disasters, failures of large equipment, and other disruptions.

Digitalization and Data Analytics: The energy sector has started digitalization and consequently generates huge volumes of data. These can be interpreted to optimize energy production, distribution, and consumption. New data analytics, machine learning, and AI are being applied to forecast energy demand and predict equipment failures to optimize operation at renewable energy plants. Digital twins of physical assets enable utilities to simulate and optimize grid performance, reinforcing energy security through the use of real-time insight into how the energy system works. Digitalization serves as the ground for making better decisions, therefore using resources more efficiently.

Blockchain and Decentralized Energy Markets: Such an emerging tool is blockchain technology, capable of creating decentralized energy markets in which energy will be directly traded from one consumer to another with no need for intermediaries. The role of blockchain in energy transactions can be that of bringing transparency, security, and efficiency to energy transactions, hence creating local energy markets and enabling P2P energy trading. Energy security will be bettered because the decentralized markets will enhance the resilience of the energy system by reducing dependency on centralized power generation and empowering consumers to take greater control of their energy supply.

CCUS - Carbon Capture, Utilization and Storage

In this aspect, the world has been trying to seek mitigation of the impacts of climate change in addition to ensuring energy security. CCUS is a critical technology that involves carbon capture from CO_2 emissions from power plants and industrial processes, storage underground, or repurposing into useful products.

CHAPTER FIVE

Geopolitical Implications of Energy Security

"Good government is not about grandiose schemes, but about paying attention to practical concerns"
John Kay

Energy security is actually not merely a technical and economic problem but, in fact, is a highly important geopolitical one that affects world stability, international relations, and the strategic tasks of states. In this, nations seek reliable and affordable supplies of energy, ensuring that the geopolitical landscape continually becomes complex and contested. The chapter deals with geopolitics and energy security, showing how energy resources shape the power dynamics between states, influence foreign policy, and contribute to conflicts and cooperation. The chapters also discuss the role of energy in emerging economies and the shifting balance of power toward a more sustainable energy world.

Energy Resources as Instruments of Power

In fact, all energy resources, and in particular oil and natural gas, have been deployed as an instrument of power and levers of influence in international relations whereby countries endowed with large energy resources often hold immense influence over those reliant on imports to meet their energy needs. This optic has born alliances, transformed energy into a coercive tool, and births energy-dependent relationships underpinning global geopolitics.

The Role of Oil: Historically, oil has been the most strategically important energy resource that has underpinned the global economy and military power. Major oil-producing countries, particularly those in the Middle East, have used their vast reserves as a means of trying to implement their will on the global stage. It was earlier said that the Organization of Petroleum Exporting Countries, abbreviated as OPEC, comprising some of the world's largest oil producers, has through the years played a significant role in controlling supply and prices through coordinated production cuts or increases. By gaining influence over global markets, OPEC influences policies in energy-importing countries and changes economies of both producing and consuming nations.

The geopolitical importance of oil is reflected in the history of conflicts and interventions in oil-rich regions. In this respect, the Middle East has been a source of geopolitical conflict to a great extent because of its colossal oil reserves. The question of the control of oil resources played a role in such conflicts as the Gulf War, in which Iraq's invasion of Kuwait was partly driven by oil-related disputes. Similarly, the influence of oil in international politics is observed in alliances formed between oil-exporting nations with influential nations, as in the case of the United States and Saudi Arabia.

Natural Gas and Pipeline Politics: Due to the abundance of supply and high demand for energy, especially in Europe and Asia, natural gas has become a major source of energy. The political geography of natural gas largely mirrors the transportation system that undergirds it - namely pipelines. For example, countries that control important natural gas pipeline transit routes can wield immense influence over supply routes and, by extension, the energy security of importing countries. In this regard, Russia-one of the world's largest natural gas producers-manipulates its natural gas supplies as a tool of political influence in Europe. Specific pipelines, such as Nord Stream 1 and 2, directly connect Russia to Germany and are controversial, building tensions within the European Union. Some member states see these pipelines as ways in which Europe will be more dependent on Russian energy and, therefore, more vulnerable to political pressure from Moscow. Parallel to this, the diversification of energy supplies through projects like the Southern Gas Corridor has a contradictory goal: to reduce Europe's reliance on Russian gas in an effort to strengthen its energy security.

Renewable Energy and Resource Nationalism: The geopolitical landscape is also being increasingly reshaped by the growing importance of renewable energy resources. In the way of transition to low-carbon energy systems, various countries' transition has raised the demand for critical minerals applied in renewable technologies such as lithium, cobalt, and rare earth elements. These metals are indispensable for manufacturing batteries, wind turbines, and solar panels, due to which they have strategically become globally important in the energy transition.

It can also be said that this is where, regarding renewable energy, countries with large critical mineral reserves are starting to flex their resources in ways similar to how oil and gas are used today. For example, China, which monopolizes the global supply of critical rare earth elements, has used its control over resources as a bargaining chip in trade disputes. The result is disruption to the potential supply of critical minerals and questions over the energy security of countries reliant on imports for their renewable energy infrastructure.

Energy Security and International Relations

Energy security has a great influence on foreign policy: energy-rich states tend to use their resources as a tool of diplomacy, while energy-importing countries pursue stable and diversified supplies through international partnerships, trade agreements, and sometimes military interventions.

Energy Alliances and Strategic Partnerships: Energy alliances and strategic partnerships are an integral part of international relations in the context of energy security-for example, the relationship between the United States and Saudi Arabia has conventionally rested on mutual interests in oil. In turn, the U.S gave security guarantees to Saudi Arabia against stable supplies of oil, creating a symbiotic relationship that shaped regional dynamics in the Middle East. Similarly, energy-importing countries like Japan and South Korea have developed strategic partnerships with energy-producing countries. Their goal would involve forging long-term supply agreements; this may be through investment in energy infrastructure, joint ventures in exploration and production, and cooperation on energy technology development. It is such alliances that are critical in maintaining energy security, especially for countries with no domestic energy resources.

Energy as an Instrument of Coercion: Energy resources can also be utilized as tools of coercion in international relationships. Countries with major energy exports may utilize supply disruptions or manipulations of the price to their favor, for political or economic ends. This tactic has come to be known as "energy weaponization. The most obvious case of energy weaponization involves the use by Russia of natural gas supplies for leverage over its neighbors, most obviously Ukraine. Numerous times, during political disputes, Russia has cut off gas supplies to Ukraine, a country through which much of Europe's imported gas is transited. These periodic cutoffs have disrupted energy supplies to Western Europe. Actions like these tend to underscore the vulnerability of energy-importing countries to the geopolitical game played by energy exporters.

Energy security now shapes trade negotiations and diplomatic engagements in the pursuit of countries with which trade agreements can be negotiated, including provisions that entail energy cooperation, such as establishment of free trade zones for energy products, removal of tariffs on renewable energy technologies, and joint development of energy infrastructure. For instance, the European Union has actively pursued negotiations on trade agreements with energy-producing countries in a bid to diversify sources of energy supply and reduce dependence on Russian gas supplies. Similarly, huge investments in energy infrastructure throughout Asia, Africa, and Europe in China's BRI project cement its clout over these regions while securing energy supplies for the rapidly growing economy.

Climate Change and Energy Diplomacy: The whole world is involved in the collective struggle with climate change, which has a serious impact on energy diplomacy, too. States are actively shaping it through multilateral negotiations or coalition building to further the development and deployment of clean energy technologies. International agreements like the Paris Agreement urge countries toward common action in reducing greenhouse gas emissions and switching toward sustainable energy systems. Therefore, given the context of climate change, energy diplomacy needs to negotiate over the goals of reducing emissions, technology transfer, and finance in developing nations to attain their objectives of energy security and climate. The example of the Green Climate Fund, established under the UNFCCC, makes clear that international cooperation is being mobilized to support energy security within the frame of climate action.

Energy Security and Conflict: Energy security has been responsible for several historical conflicts due to reasons such as competition for control over energy resources, disputes concerning the routes through which energy shall pass, and also the need to compete in having strategic reserves assured.

Conflicts over Resources: In the case of resource conflicts, scrambles or jostling of countries for access to taken valuable energy resources come about where there is a territorial or border dispute and common/ shared resources. A good example is the South China Sea, which is one of the most fraught regions of geopolitical competition in the world today because of its vast oil and natural gas reserves. Overlapping claims by several nations to parts of the South China Sea create an area where diplomatic tensions have the potential for military confrontation. China, Vietnam, and the Philippines are just three of the most disputing claimants. In like manner, the Arctic is turning into a focal point of competition for resources with the opening up of new areas due to melting ice for oil and gas exploration. Countries such as Russia, the United States, Canada, and Norway are flexing their muscles increasingly in the Arctic-frozen region of tundra, sea ice, and northern latitudes- because of its territorial claims; this action raises growing apprehension over possible conflict because open access to energy resources was hitherto not thought to be available.

Energy Infrastructure as Target: Among most of the conflicts, energy infrastructures like pipelines, refineries, and electric generation facilities are very common targets due to their strategic importance. Thus, disruption in energy infrastructures has far-reaching consequences affecting not only the countries involved in the conflict but also global energy markets. For instance, during the civil war in Syria, there have been many attacks on energy infrastructure-pipelines and oil fields-which have drastically lowered the country's oil production and contributed to the general destabilization of the region. Similarly, sabotage of oil pipelines by militant groups against Nigeria has repeatedly cut into the country's exports of oil, sending ripples through global oil prices and hurting the economy.

Energy Security and Military Strategy: Energy security is also a main consideration in military strategy. This ability to secure and safeguard energy supply lines is quite central to the operational effectiveness of military forces; in time of war, control over energy resources and infrastructure becomes a critical objective. For example, during World War II, one of the major strategic objectives of both the Allies and the Axis powers was control over the oil fields and refineries. The struggle over the oil fields of the Caucasus, North Africa, and Southeast Asia decided the course of the war. Today, the military forces in the region still consider the security of the critical energy infrastructure, such as the Strait of Hormuz, an essential chokepoint for worldwide oil transports, a key priority.

The Role of Emerging Economies in Energy Geopolitics: The recently developing economies, such as China, India, and Brazil, have become increasingly influential in world energy geopolitics due to rapid economic growth and increasing energy demand. These countries are turned from pure energy consumers into leading actors of the world energy market, which dictate international energy prospects with their consumption and investment policies as well as geopolitical strategy.

China's Energy Strategy: As the most populous country in the world and having the second largest economy, China has turned to be a protagonist in energy geopolitics. Due to rapid industrialization and urbanization, the energy demand is high, which has made the country look for ways of securing sources of energy both within and outside its territories. China has a multi-faceted approach in its energy security strategy: diversification of sources, investment in renewable technologies, and securing supplies across the globe. China has invested phenomenal amounts in energy infrastructural projects worldwide, most of them under the Belt and Road Initiative. The pan-continental infrastructure development is in the path to build and enhance trade and energy links spanning Asia, Africa, and Europe. The financial and infrastructural development of energy infrastructure like pipelines, seaports, and power plants in the partner countries secures the resources of China and strengthens geopolitical influence. Also, China plays the role of one of the leading countries in international energy transition with big investments in renewable energies, from solar panels to wind turbines, to facilitate reducing fossil fuel dependence and continuing with the aspirations regarding the climate.

Another reflection of China's rising importance in the international energy markets is through the strategic purchase of energy assets abroad. Chinese state-owned companies have purchased shares of oil and gas fields across the African, Middle East, and Latin American regions. These investments afford China access to sizeable reserves of energy, in addition to helping the country further expand its geopolitical reach and influence.

India's Energy Landscape: The rapid economic and population growth has placed a high burden on the ever-growing energy demand of the country. As an emerging major oil and gas importer, the core of India's energy policy is in developing stable and diversified supplies to fuel its growth. Focus areas of India's energy policy include the promotion of domestic production by providing appropriate incentives, diversification of energy supply, and energy efficiency.

As a developing country, India has sought various multilateral and bilateral partnerships to establish its energy needs. India has signed long-term oil supply treaties with countries such as Saudi Arabia and Iraq in the Middle East to assure similar percolations of energy resources. It is also actively participating in international energy fora and organizations like the International Energy Agency and the International Solar Alliance with the aim of developing its international standing in energy matters and cooperating in all energy-related projects.

In addition to making adequate supplies of all forms of traditional energy, India is investing in renewable energy and is trying to become one of the world leaders in clean energy technologies. While setting ambitious targets for its solar and wind energy capacity, the efforts place it well on track to achieve this goal in a pathway toward a global low-carbon energy system. In focusing on renewable energy, India is not only trying to sort out its domestic energy needs but also positioning itself as one of the key drivers in this global energy transition.

The Role of Brazil in Energy: With its rich natural resources and emerging economy, Brazil is considered to be an important player in Latin American energy geopolitics. Richly endowed with renewable energy resources, hydroelectric power is a predominant resource and accounts for an overwhelming portion of its electric generation. Therefore, the main axis of the Brazilian energy strategy is to expand the capacity base of renewable energy, develop oil and gas reserves, and enhance regional influence.

The country also employs its various energy resources in consolidating a geopolitical position within the South American continent. It has pursued regional energy integration projects, including developing cross-border electricity interconnections with countries in its vicinity, further enhancing energy security and deepening regional cooperation. Brazil's offshore oil reserves of the pre-salt fields have attracted significant interest and investment by international companies, further integrating the country into the global energy market.

Where energy resource and infrastructure development is ongoing, Brazil situates itself at the regional and global levels as an important actor in energy geopolitics. Its balancing of fossil fuel development with investment in renewable energy is representative of a broader trend across emerging economies to balance energy security and climate goals.

Shift in Balance of Power: There is a shift in the balance of power in world energy geopolitics with the rise in emerging economies. These nations, while making their presence felt in the energy sphere, are challenging established dominance and the reshaping of international energy dynamics. This increasing influence of the BRIC Countries-China, India, and Brazil-means energy is more important in world politics now than at any time in the past, as the balance in energy security shall be more participatory and multi-dimensional.

The future of global energy governance is thus set by the interaction between emerging economies and established energy powers and their involvement in international energy forums and agreements, while growth and evolution in the respective roles of the emerging economies will no doubt be more pronounced in order to influence the global energy markets, international relations, and finally the trajectory of the energy transition.

In all, emerging economies are playing a transformative role in global energy geopolitics-driven changes in energy consumption, investment, and strategic alignment. Shaping the future of energy security and helping to forge an increasingly complex and interdependent global energy landscape, emerging nations chart their own unique ways through navigating their energy needs and ambitions.

CHAPTER SIX

NAVIGATING ENERGY SECURITY AND

ENVIRONMENTAL SUSTAINABILITY

"Energy efficiency isn't just low-hanging fruit; it's fruit that is lying on the ground"

Steven Chu

In this interdependent world, the balancing challenge between energy security and environmental sustainability resonates as one of today's key preoccupations of policymakers, business, and individuals. As global energy demands are continuing to escalate, while the due effects of climate change are becoming more obvious, finding a balance between the security of reliable energy supplies and protection of the environment is crucial in underpinning prosperity and stability in the long run. This chapter looks at the intricate relationship between energy security and environmental sustainability, considers the impacts of traditional sources, highlights transitional new renewable energy, examines the role of energy efficiency and conservation, and discusses the importance of effective policy and regulatory frameworks.

The Environmental Impact of Traditional Sources of Energy

For the past century, conventional energy resources like coal, oil, and natural gas have shaped the path of industrialization and economic development. Increasingly, their use has had very serious environmental impacts, which make their long-term sustainability really doubtful.

Greenhouse Gas Emissions: Combustion of fossil fuel is indeed the single largest source of GHG emissions, especially carbon dioxide. These gases are a major cause of global warming and climatic change. Since a number of countries still rely on fossil fuels for a big part of their energy supplies, the energy sector is among the main contributors to global CO_2 emissions, and the task of reducing these emissions becomes rather acute; hence, the imperativeness of addressing this problem does not come without a switch toward cleaner energy sources and an implementation of effective carbon reduction strategies.

Air and Water Pollution: The extraction, treatment, and burning of fossil fuels all produce forms of pollution. Coal mining, for instance, can disrupt land and contaminate water due to heavy metal runoff and toxic chemicals. Drilling oil and extracting natural gas can result in oil spills and gas leaks that can decimate marine and freshwater ecosystems. In its combustion, fossil fuel emits into the atmosphere the notorious pollutants SO_2 and NO_x, further aggravating air quality problems and causing diseases in both human beings and wildlife. To eliminate these effects, other cleaner technologies must be adopted along with the implementation of firm environmental policies.

Habitat Destruction: The ecological footprint of extracting fossil fuel extends to habitat destruction and biodiversity loss. Oil drilling, mining operations, and construction of pipelines and infrastructure can disrupt ecosystems and destroy wildlife habitats. These activities not only threaten biodiversity but also have long-term ecological consequences. In the case of addressing habitat destruction, one has to consider the option of alternative sources of energy as well as reduced environmental impacts of energy infrastructure projects.

Transition to Renewable Energies

Transitioning into renewable energy systems is one of the fundamental methods for addressing the environmental impacts of traditional energy systems and attaining a sustainable energy future. The use of renewable energy sources such as solar, wind, hydro, and geothermal presents a better opportunity to reduce greenhouse gas emissions and consequently minimize environmental impacts, hence enhancing energy security.

Solar Electricity: Electricity from the sun is generated by means of photovoltaic cells or concentrating solar power systems. Operationally, it does not produce greenhouse gas emissions like those from fossil fuel-produced electricity. Advancement of solar technology has gone down in costs and up in efficiency such that solar power increasingly is feasible both on-grid and off-grid. This source of energy, therefore, gives assurance of not having environmental impacts due to a reduction in the dependency on fossil fuels and consequently reducing carbon emissions.

Wind Energy: Wind-generated electricity involves a mechanical process whereby kinetic energy from the wind is transformed into electric energy with the use of wind turbines. Wind energy, among the developing forms of renewable energy, has a very negligible effect on the environment compared to fossil fuel. There could be both onshore and offshore wind farms, each with its advantages and disadvantages. The potential of offshore wind farms is much higher, but at the same time, their possible impact on land use areas is very minimal. Careful site selection and assessment are necessary regarding possible impacts on marine ecosystems and bird populations.

Hydropower: Electricity produced by conversion of energy originally taken from flowing water. Large-scale hydropower projects, such as dams and reservoirs, are generally invariant sources that offer reliable baseload power and enhance energy security. Large-scale hydropower projects can also have significant environmental effects, including habitat disruption and the alteration of fluvial ecosystems. Alternative projects to large-scale projects involve smaller-scale hydropower projects like run-of-river systems, which have little or no ecological footprint. The establishment of sustainable management practices and thorough project planning will help strike a balance between the various benefits and challenges of hydropower.

Geothermal Energy: Geothermal energy utilizes the heat from the Earth's interior for both electricity generation and direct heating. Geothermal power plants are at a low level of greenhouse gas emissions; rather, they can provide a very reliable and stable source of energy. In general, the environmental impact on geothermal energy is relatively low. Land use change could be included, and a potential consequence might be an impact on groundwater resources. Sustainable management and technological advances are important in trying to optimize the benefits arising from geothermal energy while reducing environmental impacts.

Energy Efficiency and Conservation: Energy efficiency and conversation are crucial for reducing environmental impact and enhancing energy security. Less energy could be used for a given service or product, or overall energy use could be reduced to reduce emissions and resource depletion.

Energy Efficiency Measures: Energy efficiency improvement implies the implementation of technologies and practices that reduce energy consumption in various sectors. This would include things like efficient appliances, lighting, and HVAC that greatly reduce energy use in residential and commercial buildings. In transportation, fuel-efficient technologies and increased usage of electric and hybrid vehicles also lead to energy savings with lower associated emissions. Opportunities for improving energy efficiency in industrial processes include optimizing equipment and adopting best practices in energy use.

Energy Conservation Strategies: Energy conservation concentrates on the reduction of overall energy use through behavior modification and policy, public campaigns for awareness, incentive programs, and regulation that enforces the implementation of energy-saving practices. Energy-efficient building codes and standards, incentives to retrofit existing structures, support these efforts of conservation and decrease energy demand. Integrating renewable energy sources into the design of buildings, like solar panels and passive solar heating, further improves energy efficiency and conservation.

The Role of Smart Technologies: Smart technologies include smart meters, home energy management systems, and smart grids. They are crucial to energy efficiency and conservation. Smart meters offer real-time information on energy consumption, enabling their consumers to monitor and adjust their energy use. Home energy management systems use sensors and automation to optimize energy consumption in residential buildings. It improves the efficiency and reliability of electricity distribution by integrating smart grids with renewable energy sources, which enable demand-side management.

Policy and Regulatory Frameworks

Effective policy and regulatory frameworks are so important in aligning energy security with environmental sustainability. Governments and international organizations set standards, offer incentives, and put in place regulations that foster clean energy technologies, assure energy efficiency, and protect the environment.

In this connection, many countries have developed national policies to respond to the twin imperatives of energy security and environmental sustainability. Other tools involve renewable energy targets, carbon pricing mechanisms, and energy efficiency standards. For example, the Renewable Energy Directive adopted by the European Union sets ambitious targets for increasing the share of renewable energy in the energy mix of the EU, taken as a model policy to be emulated by other nations such as the United States and China.

Energy Storage Solutions: In the development of advanced technologies for energy storage, lithium-ion for example. IRENA provides data, policy advice, and technical support to member countries to enhance their renewable energy capacity and integrate clean energy solutions. The agency encourages international cooperation in various projects involving renewable energies, transfers technology, and best practices on the deployment of renewable energy.

World Energy Council: It is an international organization founded in 1923. Its stated purpose is to work for the sustainable supply and use of energy. The Council seeks to unite high-level energy leaders from governments, industry, and academia in discussing and resolving specific energy challenges. WEC has been involved in conducting research, conferences, and policy recommendations on energy matters. The Council was also committed to serving as a catalyst for inspiring dialogue and coordination among a wide array of stakeholders, which, in turn, contributes to integrated policy making and strategy development within the energy sector.

Emerging Trends and Future Directions

With the currently changing landscape of global energy, international institutions need to respond to new trends and challenges if they are to remain effective in promoting energy security.

Climate Change Focus: Climate change has now become part of the core priorities of international institutions. Embedding climate goals into energy policy and strategy is indispensable with respect to reaching global climate objectives and securing energy supply in the long run. For example, it is recently becoming the case that both IEA and IRENA put more emphasis on the interface between energy security and climate action by advancing policies that will support clean energy transitions and climate resilience.

Greater Emphasis on Digitalization: While the digitalization of energy systems has opened up avenues, it is equally challenging for international institutions in varied ways; thus, smart grid, digital technologies, and data analytics will be more in use, and at this juncture in time, the institutions are approached to take up issues concerning cybersecurity, data privacy, and technological integration. Hence, developing frameworks and guidelines will be of immense consequence to make the digital energy systems secure and reliable.

The Greater Role for the Emerging Economies: Emerging economies are of growing relevance to the global energy markets and international energy governance. The institutions have to engage with these countries in order to respond to their challenges and opportunities, such as how the IEA and IRENA are expanding their outreach and support to emerging economies to help them navigate energy transitions and enhance their energy security.

This can be accomplished through the strengthening of multilateral cooperation, confidence, and partnership among international institutions, governments, and private sector actors as means to underpin global energy security challenges. Collaboration enhances the effectiveness of policies, knowledge sharing can be prioritized, and support provided towards the development of innovative solutions. International collaborations, exemplified by experiences of the Global Energy Interconnection Development and Cooperation Organization (GEIDCO), possess high potential for going forward in a manner that secures energy infrastructure and integration.

Case Studies of Successful International Initiatives: Specific case studies may give a clearer idea about both the successes and challenges faced by international institutions while promoting energy security.

CEM

Clean Energy Ministerial was, in fact, one of the successful international initiatives to foster policies and technologies for clean energy that started operations in 2010; the energy ministers of major economies converge to collaborate on creating swift and innovative deployments of clean energy. Explaining the initiatives like the Super-Efficient Equipment and Appliance Deployment initiative and the Clean Energy Solutions Center by the work carried on by the CEM, one finds that efforts have been made to encourage the adoption of energy-efficient technologies along with reprimanding policy development.

The International Energy Charter: The International Energy Charter is a multilateral structure for cooperation among states and organizations on various energy issues, which was adopted in 1991 by the Energy Charter Conference. This Charter strongly focuses on energy transit, investment protection, and trade; it, therefore, has facilitated lots of dialogue and cooperation on energy security and related matters. The Energy Charter Treaty provides a legal framework for energy trade and investment, thereby contributing to stability and predictability of international energy markets.

Global CCS Institute: The Global Carbon Capture and Storage Institute probably represents one of the finest collaborations on the international plane hitherto on climate change and energy security. The Institute, upon its establishment in 2009, was charged with fostering the development and deployment of CCS technologies as a mitigating factor to GHG emissions. Through its research, advocacy, and partnerships, the Institute has supported the development of CCS projects and contributed toward global efforts toward mitigating climate change.

Recommendations for Improvement in the Role of International Institutions

The international institutions can improve their role in the field of energy security on the basis of various recommendations given as follows:

Improve Coordination and Enhance Collaboration: Improved coordination at the international level would no doubt enhance the effectiveness and efficiency of the institutions involved in putting a framework for international energy security in place. Clear structures regarding the sharing of information, acting jointly, and also implementing integrated strategies can avoid overlaps and added value, maximizing it. For example, creating joint task forces or working groups on particular energy security issues develops collaboration and harmonizes activities.

Expand Engagement with Emerging Economies: A dialogue with emerging economies, while responding to particular challenges and opportunities among those economies, can be instrumental in making international institutions more relevant and influential. Targeted support for and initiatives within the countries themselves will ensure more inclusive and effective solutions on energy security; technical assistance, capacity building, and funding support to the emerging economies will facilitate country transition into clean energy systems and improve their energy security.

Support transparency and accuracy: There should be transparency and accuracy in data reporting to help in the building of trust and credibility. International institutions should, therefore, be involved in developing mechanisms that are required for the collection, verification, and dissemination of data. The methodologies and frameworks developed and standardized for the reporting of the data would go a great way in reinforcing the dependability of the world's energy statistics for informed decisions.

Entering into Public-Private Partnerships: Public-private partnerships can provide a very fundamental basis on which to advance energy security initiatives in response to the several challenges being faced globally. Collaboration by industry stakeholders

CHAPTER SEVEN

Power and politics: The Geopolitical stakes of Global Energy Security

"Good government is not about grandiose schemes, but about paying attention to practical concern"
John Kay

Energy security has, therefore, been born into the fast-moving evolutionary process in international relations and geopolitical strategy of the present world. The geopolitical landscape is under constant transformation, especially with regard to emerging trends related to energy demands, which are pressing most countries between two difficult choices: ensuring stable energy supply versus cares for the environment. The chapter shows the interlinkage of energy security and global politics in a complex way, how energy resources, supply chains, and policies impact international relations of influence on global stability and a change in power dynamics within the world stage.

Energy Resources and Global Power Dynamics

Energy resources are one of the key factors in the international dynamics of power, shaping both the capabilities and strategies of nations. These resources happen to be unequally distributed-a fact that has developed into a complex web of interdependencies and rivalries defining international relations.

Strategic Importance of Energy Resources: Energy resources such as oil, natural gas, and coal are at the core of geopolitical strategy. The fact is that countries having a good resource base of energy grow their influence over the global market and politics. For example, the Middle East-a region with some of the world's largest oil reserves-has, for decades, been a focal point of geopolitical interest and conflict. Similarly, Russia's huge natural gas deposits give it enormous leeway over European energy security, affecting diplomatic relations and economic stability in the region.

Resource Control and Geopolitical Leverage: A country with a great amount of energy resources is able to apply geopolitical pressure due to control over accesses to such resources, allowing them to regulate world energy prices and supply security, hence affecting other economies. OPEC, or Organization of the Petroleum Exporting Countries, comprising countries with large oil reserves, has been utilizing its accumulated production capacities to influence oil prices and worldwide energy markets. This ability of adjusting production level gives them great influence in international energy matters.

Economic and Military Power: Energy supplies result in economic and military power of the nation. Countries that have abundant energy can use these resources to spur economic growth, advance technology, and further expand their military access. The wealth accrued from energy export enables the nations to invest in infrastructure, technology, and defense, and thus reinforces their geopolitical standing. Whereas countries with minimal energy supplies may wish to seek access to foreign sources through alliances or trade agreements, and sometimes will go as far as using military intervention.

Geopolitical Risks and Energy Security

Pursuit of energy security therefore carries significant geopolitical risks impacting on global stability and on international relations. These include supply disruption, market volatility, and geopolitical conflict-all requiring careful management and strategic planning.

Supply Disruptions: Energy supply disruptions can arise from a variety of sources including geopolitical conflicts, natural disasters, and technical failures. Conflicts in any energy-rich region of the world have created significant disruptions in oil supplies, leading to price volatility and economic instability. The Gulf War of 1990-1991 caused a temporary rise in oil prices due to fear of disruption in supply. Political turmoil in Venezuela and Libya has drawn global oil markets into disarray, again underlining the vulnerability of energy supplies to geopolitical events.

Market Volatility: Energy markets are especially susceptible to geopolitical shocks that could result in price volatility and economic instability. It is a fact that market volatility may result from changes in government policies, trade disputes, and conflicts. For example, the 2014 collapse in oil prices was a function of a combination of oversupply, geopolitics, and shifting demand patterns. The resulting drop in price had important repercussions for oil-producing countries and global energy markets, showing the interdependence between energy security and geopolitical stability.

Geopolitical Conflicts: The fact that the resources and supply routes of energy resources are usually confined to a few territories raises the likelihood of their becoming causes for conflict in developing countries, as was observed in the heightened tension and diplomatic confrontation the international community witnessed over the South China Sea. It is thought that the South China Sea may have a lot of oil and gas supplies in its seabed, and this has occasioned some territorial claims between China and other nations into rather contested multidimensional geopolitics. Conflicts like these may affect world energy supply chains and put stresses on international relations, which raises the imperatives for diplomatic involvement and resolution of conflicts.

Energy Security and Its Impact on Regional Stability

Energy security has strong implications for regional stability, including political relationships and economic development, and thus affects security dynamics in particular areas.

Energy Dependence and Vulnerability: Most of the countries dependent on the importation of energy are highly vulnerable to every disruption in supply and any fluctuation in energy prices. For example, many European countries rely on the Russian natural gas supplies for meeting their energy requirements; thus, these countries are highly dependent on these sources, which in turn influences regional security and even the diplomatic relations between states. Overreliance on supplies derived from just one supplier or a few sources makes the countries dependent on it vulnerable to geopolitical risks and economic instability that stir efforts toward diversification of energy supplies and further enhancement of energy resilience.

Regional Power Games: The sources of energy could be one of the causes of regional power games in which the countries contest control over strategic supply routes and prized resources. In the Caspian Sea region, for example, various rivalries have emerged due to the exploration and utilization of oil and gas resources between countries such as Azerbaijan, Kazakhstan, and Russia. The energy resources located in this region have, no doubt, become one of the main causes for shaping regional alignments and disputes that demonstrate the role of energy security in regional power dynamics.

Development and Security of Energy Infrastructure: Infrastructural security is one of the most important facets leading to regional stability. Attacks have also targeted pipelines and power plants, which consequently affects energy security and regional stability. The laying of pipelines, such as the Nord Stream pipeline connecting Russia to Europe, has been enveloped in questions of security and geopolitical implications. Thus, it would be befitting to say that the security of energy infrastructure requires coordination between governments, companies, and security agencies to avoid disruptions and maintain regional stability.

The Role of Emerging Economies in Global Energy Geopolitics

The role that is being played by these emerging economies becomes influential, as they reshape the balance of power in global energy geopolitics, having an impact on international energy markets.

China's Energy Strategy: Being the world's largest energy consumer and importer, China has a big influence on global energy geopolitics; hence, her energy strategy involves securing access to energy resources through investments, trade agreements, and geopolitical partnerships. The BRI illustrates an important pillar of the Chinese strategy for the diversification of energy security by investing in infrastructure and trade routes across Asia, Africa, and Europe. China's rise in international energy markets implies relevance for international relations and energy security since the ongoing energy policies and investments by China shape current world energy developments.

India's Energy Needs: With its fast-growing population and economy, India has gradually become a big player in world energy markets. India's growing energy needs are driving its search for new sources of energy-including investment in renewable energy-and forging partnerships with energy-producing countries. India will be expected to continue playing an increasingly influential role in global energy geopolitics, which shapes its relations with key energy suppliers and will have a profound impact on regional stability. The energy strategy for India is diversification in all energy sources, improving energy efficiency, and the development of strategic partnerships in order to ensure adequate, stable, and secure supplies of energy.

Emerging Economies and Resource Demand: Other emerging economies like Brazil, Indonesia, and Nigeria also join in, playing an increasingly important role in global energy geopolitics. With further development, these economies are likely to increasingly influence global energy policy and international relations in general, due to growing energy demand and increased resource investments with an impact on world energy markets and geopolitical landscapes. It is the emerging economies that will shape the future of global energy: investing in new technologies, pursuing energy diversification, and engaging in strategic partnerships with energy-producing countries.

Strategies for the Improvement of Energy Security and Geopolitical Risk Mitigation

The geopolitical consequence of energy security needs to be pursued multi-dimensionally: diversification of energy sources, increased international cooperation, and resilience in the energy infrastructure.

Diversification of sources- diversifying the source will help reduce dependency on a particular source or supplier. Thus, countries and companies may examine a mix between renewable and non-renewable options to achieve energy supply security and reduce geopolitical risks. New technologies like energy storage and smart grids can support the process of diversification and help in resilience to energy. The diversification of energy resources also includes development of domestic energy resources, investing in alternative sources of energy, and considering international trade agreements as ways to ensure a steady and dependable supply of energy.

International Cooperation: Cooperation at the international level is necessary to resolve global energy challenges and manage geopolitical risks. Thus, common efforts-like joint projects for energy, investments in cross-border infrastructures, and collaborative research-will further enhance energy security and lead to predictable international relations. Because of this fact, the International Energy Agency and the International Renewable Energy Agency do, indeed, have an important role in promoting cooperation and, likewise, in the advancement of global energy goals. Many disputes can be resolved through diplomatic engagement and multilateral agreements, as they address mutual grievances and foster mutually beneficial relationships.

Resilience of Infrastructure: Resilient energy infrastructure in construction and maintenance is very fundamental in managing geopolitical risks and ensuring a steady energy supply. This means investing in secure and reliable energy systems, enhancing cybersecurity, and building contingency plans in case of disruptions in supply. Through various measures protecting energy infrastructure against both physical and cyber threats-government and businesses working together can make energy infrastructure resilient to geopolitical setbacks. This is a prerequisite for maintaining energy security and regional stability.

Diplomatic Engagement: it thus necessitates superior conflict management diplomacy and the building of international cooperation, with particular emphasis on energy-related matters. For this reason, the option of entering into dialogue with key stakeholders, negotiating agreements, and participating in various international forums can be considered some of the means to resolve disputes and develop mutually advantageous relationships. While this can be avoided only by making diplomatic efforts oriented toward stability, the attending of grievances, and the promotion of common energy goals by means of trust and cooperation among nations, that way secure and sustainable energy futures are ensured.

Future Trends and Challenges in Energy Geopolitics

The main key trends and challenges in the development of the world energy configuration will define the future of energy geopolitics. First, the technological breakthroughs transforming energy demand and changing geopolitical constellations will strongly affect energy security and international relations.

Technological Change: Renewable energy technologies are the chief driver of change in the energy system and are expanding rapidly to yield far-reaching market and geopolitical consequences. Solar, wind, hydro, and geothermal power are getting increasingly cost-competitive and diffusionary, reducing dependence on fossil fuels. It might therefore present a shift in traditional power balances, wherein countries previously dependent on energy become potential exporters of renewables. For instance, countries that are endowed with sunlight or wind-wealthy nations, such as Saudi Arabia and Denmark, respectively, leverage core building blocks of their renewable solar and wind energy resources as strategic underpinnings in the diversification of their respective energy portfolios and the extension of their geopolitical influence.

Energy Storage and Grid Inventions: Technological development of energy storage technologies, including batteries and pumped hydro storage, is critical in taming intermittent renewable energy sources. Enhancements in energy storage and smart grid technologies allow for better distribution of energy, make sparing use of fossil fuel resources, and render the energy systems resilient to various degrees. Such technologies could reduce supply disruptions and enhance energy security to a level that will affect geopolitical strategies and international energy relationships.

Hydrogen Economy: Hydrogen use is slowly but surely emerging as one of the more promising alternatives to traditional fossil fuels and, thus, can hope to play an important role in future energy systems. Green hydrogen-made with the help of renewable energy sources-is a clean and versatile energy carrier for power generation, transportation, and other industrial processes. As the world starts investing in hydrogen infrastructures and technologies, the geopolitical landscape may also shift accordingly, as new players enter as key hydrogen producers and consumers.

Shifting Energy Demands

Decarbonization and Climate Goals: The increasing decarbonization of most countries' energy and the attainment of climate goals translate into a shift of energy geopolitics, with countries being more committed to a pathway of greenhouse gases reduction and transitioning towards low-carbon sources of energy. This transition to renewables affects the bottom line of traditional energy markets and could have profound implications for geopolitical alliances and rivalries inasmuch as countries that become leaders in green technologies and carbon reduction strategies may derive geopolitical leverage from this position, while those remaining reliant on fossil fuels could suffer economic and diplomatic challenges.

Rising Economies and Energy Requirements: The fast-growing economies, especially in Asia, are leading to a growing need for energy worldwide. Other emerging economies, such as China and India, are using more energy to fuel economic growth and development. Growing energy demand has implications for global energy markets and geopolitics, as these developing countries seek to improve energy supplies and invest in infrastructure, both domestically and around the world. The energy policies of these nations and their relationships with other nations will continue to play an increasingly important role in determining future geopolitics.

Evolving Geopolitical Dynamics

While energy demand keeps increasing, competition over basic resources like rare earth elements and fossil fuels is only heating up. The next step in this may be geopolitical rivalries over resource-rich regions, leading to greater tensions and conflicts. Countries might opt to secure their interests in resources through strategic investments, partnerships, or even military actions. For example, the Arctic, with its unexploited supplies of oil and gas, is turning into a hotbed of geopolitical rivalry as the melting of ices opens new shipping routes and gives access to resources. Energy Transition and Geopolitical Shifts On the one hand, transition from fossil fuels to renewable sources of energy promises a change in geopolitical dynamics.

Consequently, countries that can make this transition with a much more solid renewable energy sector may be strategically positioned to have more prosperous economies and political regimes than those unable to adapt to this energy transition. In such cases, new alliances and rivalries could emerge over questions of energy technologies and sustainability goals. Undeniably, the transition of global energy toward the future will definitely require reassessment of conventional geopolitical relationships in light of a new strategy for energy security management. Geopolitical risks and cybersecurity: Energy systems are going to be more interconnected and digital; therefore, cybersecurity will become critical in energy geopolitics.

Cyberattacks on energy infrastructure may result in the disruption of supply chains, damage to equipment, and impairment of national security. Countries need to make investments in robust cybersecurity measures and international cooperation in protecting their energy systems against cyber threats. The emergence of digital technologies and the further complication of energy infrastructure are bound to underpin the need for new approaches in the management of geopolitical risks and in ensuring the security related to energy assets.

Conclusion

Energy security strongly complements the geopolitics mowing global power structures, regional stability, and international relations.

While navigating through various complexities associated with securing supplies of energy and catering to their related environmental challenges, they are concurrently obliged to take into consideration geopolitical implications associated with their strategies on energy. By grasping the role of energy resources in geopolitical strategy, managing geopolitical risks, and leveraging international cooperation, countries can improve their energy security while contributing to stability and sustainability globally. The ability to navigate such geopolitical challenges in a perpetually changing global energy landscape will be fundamental to achieving a secure and sustainable energy future.

CHAPTER EIGHT

ENERGY SECURITY AND ENVIRONMENTAL SUSTAINABILITY

"In an interconnected world, we all depend on each other for energy security".

Fatih Birol

The twin challenges facing the world are those of energy security and environmental sustainability. In the search for a secure and reliable supply of energy, a growing intersection with environmental protection and tackling climate change is being pursued. This chapter explores in depth the complex interlinking of energy security and environmental sustainability; the challenging development of energy policies, technological innovation, and international agreements that can put these often-conflicting goals together.

Energy Security and Environmental Sustainability:
One cannot overstress how much their interrelationship has taken on a competitive objectiveness since they are interconnected. So, achieving energy security with minimum environmental impacts requires a nuanced approach that considers both short-run needs and long-term goals.

Energy Security and Environmental Trade-offs: The pursuit of energy security does indeed have a historical precedence compromised on the altar of short-term economic and strategic benefit. Reliance on fossil fuels, for example, has provided a stable and relatively affordable supply of energy, while at the same time it has contributed to tremendous environmental degradation and climate change. Therefore, balancing these trade-offs involves adopting strategies that improve energy security with the least environmental impacts, such as transitioning toward cleaner sources of energy and the realization of energy efficiency.

Actions based on integrated policy are required: policymakers need to think of and implement integrated policy actions in the search for compatibility between energy security and environmental sustainability. These might include renewables promotion policies, enhancement of energy efficiency, and reduction of greenhouse gas emissions with regard to both energy and environmental goals. In this manner, integrated policies can contribute to a more robust energy system with reduced environmental burdens, insuring consistency between energy security and sustainability goals.

Technological Advancement for Sustainable Energy Security
Technological development is imperative for both energy security and environmental sustainability. With the development of new technologies, the production, storage, and efficiency of energy are being rethought amid a changing dynamic in the energy sector that will help find solutions to these two challenges.

It involves development and deployment of renewable energy technologies, the centerpiece of achieving environmental sustainability coupled with the security of energy supplies. Solar, wind, hydro, and geothermal energy provide clean and sustainable alternatives to fossil fuels. These technologies reduce greenhouse gas emissions and dependence on imported energy resources, thereby contributing to both the environmental and energy security goals of being at or below 1990 levels by 2010. This, of course, means that the increased capacity of solar and wind power in various parts of the world has reduced emissions and raised the level of energy independence.

Energy Storage Solutions Energy storage technologies involve advanced batteries and pumped hydro storage, which play a major role in the integration of renewable energy into the grid. It is the storage solutions that manage the intermittency in the energy sources in order to ensure that energy supply is constant and reliable. Innovation in energy storage can also contribute to strengthening resilience in the grid in a way that contributes to reducing dependence on fossil fuel supplies while shifting towards a more sustainable energy system.

Energy Efficiency: A key strategy in improving environmental impacts while enhancing energy security involves improving energy efficiency. These identify energy-efficient technologies and practices that reduce consumption of energy and greenhouse gas emissions while lowering costs and increasing energy reliability. Energy efficiency measures can be implemented at large scales in sectors such as buildings, transportation, and industry to contribute significantly toward sustainability and security objectives.

Policy Frameworks and International Agreements

What will align energy security with environmental sustainability are effective policy frameworks and international agreements. It hence requires several governments, international organizations, and stakeholders to cooperate in the formulation and implementation of policies addressing interconnected challenges.

National Policies Energy: National energy policies have taken center stage in framing energy security and environmental sustainability. Hitherto, therefore, governments are called to set clear and ambitious objectives regarding greenhouse gas emissions reductions, the advancement of renewable energies, and the enhancement of energy efficiency, but the policies regarding carbon pricing, subsidies for renewable energies, and regulations on energy consumption will indeed drive the transition toward a more sustainable and secure energy system. For example, the Green Deal presents a carbon neutrality roadmap, increased energy security, among other policy recommendations by the European Union.

International Climate Agreements International agreements on climate, such as the Paris Agreement, represent one global drive for the reasons of climate change and environmental sustainability. They set targets to decrease greenhouse gas emissions and encourage countries to adopt policies that would be in favor of transition toward cleaner sources of energy. International cooperation and commitment to climate goals are key elements in global alignment of energy security with environmental sustainability.

Energy Cooperation and Partnership: Collaborative efforts by countries and organizations improve energy security and environmental sustainability. Through partnerships in both bilateral and multilateral contexts, sharing knowledge and resources in technology will be facilitated to support the development of sustainable energy solutions. International cooperation in enhancing renewable energy technologies and best practices is fostered through initiatives like the Clean Energy Ministerial and IRENA.

Case Studies: Balancing Energy Security and Sustainability

Real-life case studies offer good insight into how energy security is being balanced with environmental sustainability both at the country and organizational levels.

Energiewende, Germany's Energy Transition: Germany's transition is one of the important examples of a country that aspires to realize energy security in a sustainable environment. The Energiewende targets nuclear power reduction, increasing the contribution of renewable energies, and increasing energy efficiency. On a more optimistic note, Germany has achieved much in expanding its renewable energy capacity, cutting greenhouse gas emissions, and improving energy security. Its experience represents both the challenges and opportunities that lie ahead for the transition to a sustainable energy system while energy security is ensured.

Expanding Renewable Energy in China: China has emerged as a leading player in the realm of renewable energy development worldwide due to heavy investments in solar, wind, and hydro projects. In addition to meeting its own commitment to environmental sustainability, the country's renewable energy policies and investment have provided fodder for 'energy security' by reducing the country's reliance on imported fossil fuels. The Chinese experience well illustrates how mega-scale renewable energy projects can meet goals of sustainability and security alike.

The balancing of energy security and environmental sustainability is also exemplified by California's policy on clean energy. The state has adopted ambitious policies to advance renewable energy, enhance efficiency, and cut down greenhouse gas emissions. As such, California's commitment to clean energy has given rise to a significant expansion of solar and wind power, improved energy reliability, and reduced environmental impacts. The state experience also underlines that fully integrated policy frameworks along with technological innovation are vital to meet the goals relating to sustainability and security.

Challenges and Opportunities Likely to Emerge: Energy security and environmental sustainability will continue to unfold both challenges and opportunities through the global evolution of energy scenarios, amidst which addressing such challenges requires continuous innovation, collaboration, and commitment over long-term objectives.

Hard-to-abate sectors are those driven by heavy industry and transportation, whose operations pose challenges for reduction in greenhouse gas emissions due to their dependence on fossil fuels. As such, decarbonization of the hard-to-abate sectors will be about developing new technologies, including carbon capture and storage, or alternative fuels. In fact, their technological advances pave a way to achieve environmental sustainability with energy security.

Energy Transition and Social Equity: The transition towards a sustainable energy system needs to be socially equitable in the sense that there is an appropriate distribution of benefits accrued from clean energy, with the purpose of addressing energy access and affordability. The key factors for a just and equitable energy transition will include issues of energy access and affordability along with job displacement. Meanwhile, policymakers and stakeholders have to collaborate on these social dimensions and ensure that the transition serves all communities.

Adaptation to the Impacts of Climate: With more and more evident impacts of climate change, countries will have to adapt their energy systems to withstand extreme weather events and other climate-related challenges. Creating resilient energy infrastructures along with developing strategies for managing climate risks will be quite important in attaining energy security with environmental sustainability.

Moving International Cooperation Ahead: The future of international cooperation will also remain central in the quest for lasting energy and environmental solutions on a global scale. While the process is slower, stronger partnerships, sharing of best practices, and coordination across borders would eventually help hasten progress toward sustainable energy solutions. Global cooperation in research, technology development, and policy implementation holds the key to a secure and sustainable energy future.

Conclusion

Energy security and environmental sustainability can be considered Siamese twins, for they have to be framed together for stability in the case of an energy resilient future. It requires technological innovation, appropriate policy frameworks, and international cooperation to align these objectives toward the path of attaining a sustainable energy system. In balancing their needs of reliable energy supplies with the imperatives to protect the environment, nations can create a more secure and sustainable energy future. Ahead lies a path that will need ongoing commitment, innovation, and cooperation to get through the complex challenges and opportunities.

CHAPTER NINE

THE ECONOMIC IMPLICATIONS OF ENERGY SECURITY

"The price of power is rising because the price of powerlessness is unbearable".
Robert Kuttner

Energy security is a question not only of supply and policy but also largely an economic one: Security and accessibility of energy sources stemming from both physical and political sources of instability do affect economic growth, investment, and international trade. In this chapter, the complex set of economic consequences of energy security are approached by discussing how energy supply disruptions, market dynamics, and policy decisions affect the world economy.

Energy Security and Economic Growth: Energy security is also fundamental to economic growth in that the cost and reliability of energy supplies impact industrial production, transportation, and daily life.

The cost of energy is a principal driver of economic activities, and therefore, energy-intensive industries, such as manufacturing and transportation, depend heavily on stable and inexpensive fuel supplies. In this respect, energy-intensive industries usually tend to be highly sensitive to fluctuation in energy prices, which in turn significantly impacts their production costs, profitability, and competitiveness. A typical instance is that of increased oil prices that lead to rising transport and production costs, generalizing the cost structure for businesses and consumers in general. On the other hand, stable and lower energy prices spur economic growth by reducing the cost of operation and increasing consumer expenditure.

Investment and Industrial Development: A secure and predictable energy environment attracts investment in industry and development. Energy security minimizes the risk of disruptions in the supply of energy; hence, it is more attractive for industry players to invest in new projects and technologies. For instance, secure sources of energy would enable the establishment of energy-intensive industries such as petrochemical and steel industries, that are vital to economic development and job creation. In contrast, insecure sources of energy discourage investment and industrialization and are capable of translating into no economic growth.

Economic Resilience: Energy security enhances the resilience of an economy by guaranteeing a sufficient supply of energy during an emergency or disruption. With respect to energy sources and infrastructure, a diversified country is much better positioned to absorb shocks caused by geopolitical conflicts or natural disasters that may affect energy supplies. Resilience limits the impact on economic stability and prevents severe disruptions in economic activities. Countries that have invested in diversified energy sources and pursued energy storage and enhancement of grid infrastructure are in a better position to handle disruptions in their supply and maintain economic stability.

Impact of Energy Market Dynamics: Energy market dynamics include price volatility, supply-demand imbalance, and market concentration-things that have enormous economic impacts.

Price volatility: Energy markets are susceptible to price volatility caused by, among other factors, geopolitical tension, disruption in supply, and changes in demand. The fluctuations in energy prices come with a broad set of economic implications, which include inflation, decreased consumer spending, and investment patterns; for example, abrupt hikes in oil prices make transportation and production more expensive, hence feeding into inflationary pressures which may retard growth. Conversely, prolonged low energy prices boost economic activities since companies and consumers will have reduced expenses.

Supply and Demand Imbalances: It is clear that the imbalance between supply and demand for energy can affect the economy's stability and growth. Short-term shortages could lead to inflationary prices and economic disruption, while over-supply could ultimately see prices fall and financial problems for the producers. For instance, the collapse of oil prices in 2014 was due to oversupply and weak demand, which cost the countries and companies dependent on oil production greatly. As such, any imbalance in demand and supply should be managed through close monitoring and forecasting so as to keep energy markets stable and favorable for economic growth.

Market Concentration and Competition One can define market concentration of energy as when only a few players hold the major market share, and in turn, affect market dynamics and economic consequences of any particular market. On the flip side, it leads to less competition, higher prices, and reduced innovation. For example, the few big oil companies that control the global oil market influence global pricing and supply decisions on oil; since they have impacts on the overall economy, it will be comforting, diversifying energy sources to reduce risks of market concentration. The more competitive the grounds for energy markets are, the more stable they will be.

Geopolitical Risks and Economic Stability

The geopolitical risks related to energy security may have a deep impact on economic stability and international trade.

Supply Chain Disruption in Energy: There is a potential for geopolitical conflict, sanctions, and trade disputes that might disrupt the supplies of energy, while their ramifications affect economic stability. Conflicts within the energy-rich Middle East region disrupt oil and gas supplies and create spikes in prices and economic instability. Similarly, trade disputes and sanctions result in affecting the flow of energy resources from producers to consumers. However, geopolitical risk management involves immense diplomatic efforts complemented with strategic reserves and diversified sources of supply that buffer such potential economic effects resulting from disruptions in supplies.

Energy Export Dependence: Most countries are highly dependent on energy exports, which exposes them to economic risks in line with changes in global energy markets. Economies depending on exports tend to be susceptible to price volatility and changes in demand that influence their economic stability and fiscal health. For example, countries such as Venezuela and Nigeria became so dependent on oil exports that their economies suffered when oil prices fluctuated or demand slowed. As stated here, a country can reduce dependence on energy export and therefore diversity their economy to attain less vulnerability to the risks at hand.

Energy Infrastructure Vulnerability: Energy infrastructures are vital to economic stability; hence, it is potentially vulnerable to physical attack, natural disasters, and cyber threats along with pipelines, refineries, and power plants. Actually, the destruction of energy infrastructure usually means colossal economic losses: loss of production, repair expenses, and supply chain delays. In this respect, robust and secure energy infrastructures have a substantial role in maintaining economic stability and cushioning the impact of such disruption.

Policy Responses and Economic Strategies

The role that policymakers play in the economic impact of energy security is seen with a variety of strategies and interventions.

Energy Diversification: Diversification of energy sources and supply routes reduces overdependence on a particular source or on any single supplier to enhance energy security and, simultaneously, economic stability. Renewable energy development, energy efficiency, and alternative fuel policies give added credence to diversified energy sources. This might mean diversification into renewable energy resources such as solar and wind power, which reduces the dependence on fossil fuels and all their negative impacts on the environment, thereby generating a resilient and sustainable energy system.

Strategic Reserves: Stockpiling strategic reserves of the most valuable fuels, such as oil and gas, may help to soften the economic impact of supply disruptions. Strategic reserves allow for a buffer in the event of temporary shortfalls in supply and spikes in prices to stabilize markets and protect economic interests. Accordingly, countries with strategic reserves will be in a better position to handle disruptions in supplies and, therefore, decrease the associated economic effects of abrupt changes in energy markets.

Energy-Efficiency Policies: These sets of energy efficiency policies would go a long way in actual reduction of energy consumption, cost-cutting, and enhancement of economic resilience. The measures on energy efficiency can be achieved through improving the quality of building insulation, industrial processes, and offering incentives toward the use of energy-efficient appliances. Energy bills can be reduced, as well as a decrease in the general demand for energy. Energy efficiency policies contribute to economic stability through cost reduction and increased business competitiveness.

Economic Diversification: This reduction in dependence on energy exports and the diversification of the economy can further give a boost to economic resiliency and stability since most countries dependent upon heavy energy revenues could suffer economic problems during periods of low or lacking prices or demand. Further development of other sectors of the economy, such as technology, manufacturing, and services, could minimize risks for such an economic foundation to become more properly balanced.

The Role of International Cooperation

International cooperation is essential in addressing global economic implications regarding energy security, and efforts of collaboration could improve energy stability, investment, and sustainable development.

Global Energy Governance: International organizations and fora provide vital roles in shaping global energy policies and facilitating cooperation on energy issues. Organizations like the International Energy Agency (IEA), the International Renewable Energy Agency (IRENA), and the World Energy Council (WEC) offer dialogue, sharing of best practice, and development of policies toward global energy challenges. These platforms provide a forum for countries to cooperate on energy security, market stability, and sustainable development.

Cross-Border Energy Projects: Interconnectors and pipelines for cross-border energy projects contribute towards greater energy security and provide a boost to the growth of an economy. These are cooperative projects that, by integrating the grids, share resources and come together to integrate markets, hence providing more reliability with lower energy costs. One of the most ambitious cross-border projects is the North Sea Wind Power Hub, which is envisaged to interlink several countries' wind power capabilities to promote stronger energy security and regional economic development.

Investment and Technology Transfer: International cooperation on issues of investment and technology transfer will help in the development and deployment of advanced energy technologies. The deliberate collaborative efforts in investing in research, sharing knowledge, and transferring technology will accelerate progress toward sustainable energy solutions and enhance economic growth. Innovation and technology development collaborative countries shall have shared expertise and resources that drive economic and environmental progress.

The Future of Trends and Economic Implications: The future trends will be stating the economic implications of energy security that influence world markets and the stability of economies.

Energy Transition Costs: The transition to a low-carbon energy system does not come without associated costs-investments in renewable energy infrastructure, technological development, and policy implementation. These costs can be associated with consequences on economic growth and investment patterns; it will be paramount that such economic costs of transition are balanced with the benefits to achieve the sustainability of the energy system in the long term, together with environmental goals.

Decentralization of Energy Production: The trend of dispersion regarding energy production through solar rooftop panels and local energy systems continuously promotes changes in traditional energy markets and economic structures. It increases resilience by reducing reliance on a few centralized power plants, but it can also interfere with established market dynamics in the search for new regulatory frameworks. Adaptation to these new potential changes will then be imperative for economic stability and fostering innovation.

Digitalization and Smart Grids: It is believed that digitalization of energy systems and the development of smart grids create new opportunities for better energy efficiency, cost reduction, and economic stability. Digital technologies optimize the management of energy consumption, allow demand response, and enhance the reliability of the grid. Later on, adaptation to shifting energy demands and growth in the economy will involve the adoption of measures related to digitalization and smart grids.

Climate Change and Economic Risks: The impacts of climate change will be economic in nature-through increased frequency and severity of extreme weather events and altered patterns of resource availability related to energy security-and learning to adapt to such risks and developing strategies for the management of climate-related challenges will thus become an imperative for stability within an economic context and for the assurance of energy security in the future.

Conclusion

Energy security is an issue of major and multi-dimensional economic implications, impacting growth, investment, and stability in global markets. The dynamics of energy supply and market conditions are closely interlinked with policy considerations.

CHAPTER TEN

ENERGY SECURITY IN A CHANGING GLOBAL LANDSCAPE

"The price of power is rising because the price of powerlessness is unbearable"
Robert Kuttner

Energy stands at the fulcrum of a dramatically changing world driven by rapid technological evolution, changing economic dynamics, and an evolving geopolitical reality. Drawing on such dynamics, this chapter considers what such changes portend for the changing concept of energy security and what that means for nations, businesses, and individuals across the world. It reviews selected developments, emergent trends, and strategic responses necessary to navigate this new environment.

Technological Advancements and Energy Security: Technical alteration of renewable energy technologies has transformed the energy security landscape. Solar and wind now boast increased efficiencies and cost-effectiveness, competing with traditional fossil fuels. This in turn has contributed to a wider variety in the energy mix and less reliance on imported sources of fossil fuel, hence better energy security. Particularly, the recent fall in the cost of solar PV has gone as low as to make the source competitive with other conventional sources of energy, hence enabling more countries to expand their usage in their respective energy systems.

Energy Storage Solutions: Development of advanced energy storage technologies, including lithium-ion batteries and pumped hydro storage, becomes important in handling intermittency related to variable renewable energy sources. The energy storage systems are capable of storing excess energy produced during high production periods and releasing it at will when production is low, hence maintaining grid stability under reliable energy supply. This facility further enhances energy security and reduces the risks associated with the variability of renewable energy sources.

Geopolitical Shifts and Energy Security

The Growing Influence of Emerging Economies: New global players, in particular those from Asia, are gaining increasing importance in global energy markets. Large and growing energy demand by countries like China and India influences global energy prices, investment patterns, and supply chains. These countries are making huge investments in fields such as renewable energies, efficiency in energy use, and infrastructural development to achieve the requirements for economic growth and energy security. This gradually changes the whole global outlook on energy dynamics and affects traditional powers in energy.

Energy Resource Nationalism: The rise in energy resource nationalism, where countries give greater priority to control of their natural resources for economic and strategic purposes, is another factor that molds global energy security. Large energy reserve countries such as Russia and nations within the Middle East often employ their resources as a leverage tool to influence regional and international political and economic developments. This could lead to geopolitical tension and disruption of supply, and shifting patterns of trade in energy around the world. For instance, the fact that Russia controls the natural gas supplies to Europe has been a source of geopolitical leverage and tension. Changing Energy Trade Routes: Shifts in the routes of energy trade are impacting world energy security. New pipelines, shipping routes, and infrastructure projects are altering the flow of energy resources and the geopolitical landscape. The development of the Trans-Pacific Pipeline and the Northern Sea Route through the Arctic, for example, creates new supply routes and alters the dynamics of world trade. With these changes, the security of the supply chain may be influenced, leading to a reshaping of geopolitical alignments.

Energy Security: The Economic Dimensions Investment in Energy Infrastructure: Recognition of the need for robust and secure energy infrastructure is driving significant investment in both developed and developing economies. These will specifically call for infrastructure investments in pipelines, refineries, and power plants that can adequately ensure reliable energy supplies to support economic growth. The nature of this investment is so capital-intensive that the development of resources within countries with limited resources may turn out to be rather challenging. Handling such challenges often requires public-private partnerships and international cooperation that guarantee the development of secure and resilient energy infrastructure.

Price of Energy: Energy prices directly relate to the economy's stability and growth. Energy price volatility has an effect on production costs and, in turn, on inflation and consumer expenditure. For instance, high oil prices inflate the cost of transportations and manufacturing, passing those costs back to the consumer in the form of higher prices. On the other hand, lower energy prices also benefit the consumers and businesses through cost reductions. However, it would have its impacts on the financial health of energy-producing countries and their companies. Hence, volatility in price and stable energy prices are needed to be managed to maintain economic stability.

Energy Security and Creation of Jobs: A more diversified and sustainable energy system in transition may result in new economic opportunities and jobs. Employment is being created in manufacturing, installation and maintenance by the growth of renewable energy industries, energy efficiency technologies and smart grid solutions. New job opportunities are arising from the engineering, construction, and project management activities of the manufacturing and installation of solar and wind energy projects, among others. The economic growth which these job opportunities bring about will support the transition to a more secure and sustainable energy system.

Policy and Regulatory Responses

National energy policies: There is an urgent requirement to have effective national policies on energy that address the dynamic global situation. The government should frame and implement policies to promote energy diversification, deployment of renewable energy, and efficient energy use. Renewable energy standards, energy efficiency mandates, and investment incentives are some of the policies that may accelerate the movement toward a secure and sustainable energy system. In other words, feed-in tariffs and tax credits can be pivotal in attracting investment in renewable energy projects and accelerating the development of clean energy technologies.

International Agreements and Cooperation: International agreements and cooperation are of paramount importance while responding to global challenges related to energy security. International agreements on the Paris Agreement and the Energy Charter create frames for countries to cooperate on climate action, energy transition, and security. Energy research, technology development, and the implementation of energy policy are areas of international cooperation in finding common challenges and goals. Examples include joint work on CCS technology for faster progress towards the goal of reducing greenhouse gas emissions and increasing energy security.

Regulatory Frameworks for Deploying Emerging Technologies: Researching, developing, and deploying emerging technologies that include smart grids and energy storage require enabling regulatory frameworks. In essence, there is a need for governments to create regulations that would encourage innovations yet protect citizens against potential dangers and other risks emanating from emergent technologies. For example, regulatory approaches that guarantee interoperability, data privacy, and cybersecurity have become so crucial for the deployment of smart grid technologies. In this light, well-defined and consistent regulatory frameworks will go a long way in helping the growth of innovative technologies and improving energy security.

Energy and Environmental Considerations

Balancing Energy and Environmental Goals: This would be one approach to merging energy security with the concerns for the environment. It will help reduce environmental impact and increase energy security by moving towards renewable energy sources, combined with the application of energy efficiency measures. It is equally important to consider the fact that energy production and infrastructure development may have environmental impacts. For example, the production of renewable technologies like solar panels and wind turbines involves extraction and manufacturing processes that are potentially harmful to the environment. Therefore, the trade-offs between the two become crucial to achieve both the objectives of energy and environmental protection.

Climate Change and Energy Security: The impact of climate change impinges on energy security in various ways: extreme weather events, shifting resource supplies, and disruption to energy infrastructure. Adaptation to climate change and the development of resilient energy systems are, therefore, fundamental to enhancing energy security in the context of unpredictable environmental conditions. A typical example of investment in resilient infrastructure includes flood-resistant power plants and heat-resistant pipelines that reduce extreme weather impacts and ensure a consistent supply of energy.

The pursuit of energy security shall therefore be in congruence with the overarching sustainable development imperatives, which include poverty reduction, economic development, and environmental protection. Energy security considerations integrated within the sustainable development strategies can spur fair access to energy, economic growth, and environmental protection. For instance, access to clean energy technologies in the developing regions through certain programs reduces poverty and improves standards of living while improving energy security.

Future Directions and Strategic Responses

Adaptation to Technological Changes: The rapid tempo of technological change requires continuous adaptation and strategic planning. Governments, businesses, and individuals should stay informed about various emerging technologies and their potential implications for energy security. Indeed, new approaches to integrating new technologies-such as advanced energy storage and digital grid solutions-can lead to more energy security and addressing evolving challenges. Investment in research and development is needed, through fostering innovation and encouraging collaboration among relevant stakeholders, if one is to stay ahead in technological advancement.

International Collaboration: Furthering international partnerships and cooperation on global energy security challenges and issues of sustainable development. Joint/crucial efforts relating to energy research, technology development, and policy implementation together can enhance global energy security and contribute to the shared goals of sustainable development. Reinforcing international collaboration by sharing best practices and coordinating cross-border efforts will be very necessary as shared challenges are tackled toward a more secure and sustainable energy future.

Finally, increasing resilience and adaptability: the significant ways for countries to cope with uncertainties and new emerging challenges involve investments in solid infrastructure, diversification of energy sources, and contingency planning-things that can help build resilience and ensure that adequate supplies of energy are available. Applying flexible and adaptive strategies, such as scenario planning and risk management, will enable organizations and governments to deal with the intricacies of a changing global energy landscape.

Conclusion

Technological advances, geostrategic change, economic imperatives, and ecological constraints mark the changing contours of energy security. The realities of this new landscape can be confronted only if there is an interrelated understanding of the play of forces that is shaping energy security, bolstered by strategic responses to the emerging challenges. Nations, companies, and people will be in a better position to respond to the challenges ahead and construct the path to a more secure and sustainable energy future if they continue to be well-informed, if they can collaborate more effectively, and if they remain innovative. The way forward would need to balance energy security imperatives with considerations for environmental impacts, technology changes, and geopolitical risks.

CHAPTER ELEVEN

Energy Security and the Role of International Institutions

"In an interconnected world, we all depend on each other for energy security"

Fatih Birol

The ever-increasing complexity and interdependence of energy security in the world today find a parallel in the development of international institutions influential in the shaping of policies, fostering cooperation, and addressing challenges at the global level. The chapter will address the established and newly emerging role of international institutions in promoting energy security, scrutinizing functions, successes, challenges, and the evolving landscape of international energy governance.

Overview of International Institutions in Energy Security

International institutions are organizations that bring nations together to cooperate and coordinate on key global issues. They provide dialogue platforms, frameworks for cooperation, and address cross-border energy challenges in energy security matters.

International Energy Agency - IEA: Formed in response to the oil crisis in 1974, the IEA is a relevant institution for energy security. The main objective of the IEA is to achieve, through response mechanisms, sharing of data, and policy recommendations, energy security for its member countries. The IEA carries out detailed analysis with respect to energy markets and provides advice on policies while facilitating combined effort in tackling global energy-related challenges. These include the IEA's emergency response mechanisms put in place for managing disruptions in oil supply and to stabilize markets, such as the Coordinated Emergency Response Measures.

International Renewable Energy Agency: IRENA was founded in 2009. It advances the adoption of renewable energy technologies. It supports the transformation of the energy system towards sustainability. Basically, IRENA works on providing data, policy advice, and technical support to its members for better renewable energy capacity and integration of clean energy solutions. The agency advances international cooperation with respect to various renewable energy projects, such as technology transfer and dissemination of best practices for renewable energy deployment.

World Energy Council: The WEC is an organization created in 1923 and is responsible for fostering the supply and use of energy in a sustainable way. The Council operates as a unified instrument for cooperation among governments and industries, coupled with academic institutions sharing mutual concerns on issues related to energy. Some of the activities undertaken by WEC include energy research and conferences, which provide the basis for policy recommendations on energy concerns. Because of this emphasis on dialogue and cooperation between stakeholders, the WEC enhances the creation of integrated energy policies and strategies.

Success and impacts of international institutions

International institutions have contributed much to energy security through their various functions and programs.

Energy Cooperation Promotion: Among the most valued creations of these international institutions is the way they advance and stimulate cooperation among nations. Energy dialogues under the leadership of IEA and renewable energy forums organized by IRENA are examples in which countries participate in developing energy policy, sharing best experiences, discussing common challenges. For example, the annual report Energy Technology Perspectives by IEA conveys insights from emerging technologies and policy trends, enabling member countries to accordingly align their strategies and invest in solutions that are innovative.

Supporting Energy: International institutions have been instrumental in supporting Transitions of the global toward cleaner and more sustainable energy systems. IRENA's efforts in increasing renewable energy adoption and WEC's focus on sustainable energy practices quickened the deployment of clean energy technologies and drove investment in renewable energy projects. Various countries and international organizations created the Clean Energy Ministerial, a ministerial forum for coordinating and promoting clean energy technologies and policies among its members.

Improvement of Data and Analysis: In addition, international institutions have achieved many impressive milestones within the domain of the provision of more accurate and detailed data. Energy Market Reports by IEA, analysts' views of Energy Technology Perspectives give a much-needed insight relevant to energy market trends, policy evolvements, and technological advancements. Such reports support the empowerment of policy planners to arrive at decisions and strategies in respect of informed investment towards improving energy security.

Emergency Response: International institutions have devised several mechanisms for crisis management in case energy supplies get disrupted. The strategic reserves and mechanisms for systematic response that IEA sets up have been vital in the stabilization of supplies when disruptions occur in oil markets. For instance, the coordinated release of the strategic oil reserves during the Libyan conflict in 2011 decreased the dent on the supply in the oil markets worldwide.

Challenges and Limitations of International Institutions

Notwithstanding the various successes, international institutions face challenges and limitations pertaining to promoting energy security at multiple points.

Geopolitical tensions could weaken international institutions through geopolitical tensions and tussles of national interests. Countries might consequently place their respective imperatives of energy security over the collective objectives of international cooperation, a process fraught with controversy and consensus decision-making difficulties. For example, the divergent policies and interests on energy matters by its member countries occasionally lead the IEA to face problems in formulating unified strategies and responses.

Resource Impoverishment: Most international institutions lack the resources and budget that would advance or implement their roles as set. Budgetary constraints could reduce the scale and nature of activities conducted, limit the scope of research, and the capacity to provide support for its member countries. IRENA, for instance, by virtue of its meager budget, might be constrained from offering comprehensive support and technical assistance to all member countries, especially the less resourced.

Data Accuracy and Transparency: International organizations must ensure data accuracy and transparency. Incomplete or inaccurate data reduces the credibility of their analyses and recommendations. For example, several countries report discrepancies in the production and consumption of energy; such disparities raise questions about the reliability of global energy statistics and further affect policy decisions.

Coordinating between international institutions and ensuring effective integration of their activities can prove a challenge. Overlapping mandates, dissimilar priorities, and lack of communication between institutions could result in inefficiencies and duplication of efforts. For example, while coordinating the two Renewable Energy Initiatives under IRENA and the WEC, a harmonized goals and activity alignment should be in place, as much as possible, to prevent redundancy and duplication of efforts.

Emerging Trends and Future Directions

With the continuous evolution of the global energy landscape, international institutions will have to be more adaptive and innovative in responding to new trends and challenges if they are to remain effective promoters of energy security.

Climate Change Focus: The thoughts of international institutions today come to the fore on climate change. It is beyond doubt that achieving global climate targets and ensuring long-term energy security is inseparable from incorporating climate goals into energy policies and strategies. From both the IEA and IRENA, an increasing concentration focuses on the nexus of energy security and climate action through the promotion of policies that will support both clean energy transitions and climate resilience.

More Emphasis on Digitalization: Digitalization of energy systems involves opportunities as well as challenges for international institutions. While smart grids, digital technologies, and data analytics are growing, so is the focus of institutions on cybersecurity, data privacy, and technological integration issues. The framework building and guidelines over digital energy systems will be the key to their security and reliability.

More critical would be the role of the emerging economies in international energy markets and global energy governance. Such challenges and opportunities, coming to their doorstep, require the need for institutions to engage with them: the IEA and IRENA are widening their outreach and support to emerging economies in support of energy transitions and improvements in their energy security.

The same development makes it imperative to reinforce multilateral cooperation in matching international institutions, governments, and private sector actors so that there is a voice for the people in meeting global energy security challenges. Sharing knowledge through collaborative efforts allows policies to be more effective, sharing knowledge, and providing innovative solutions. The progress of initiatives like GEIDCO underlines the potential international collaboration can bring about updates on energy infrastructure and integration.

Case Studies of Successful International Initiatives: It is from an analysis of case studies that one gets to understand the successes or even failures of international institutions while trying to enhance energy security.

Clean Energy Ministerial: CEM is another successful international initiative that works in the direction of promoting clean energy technologies and policies. Formed in 2010, it brings together the ministers of energy of the major economies with the objective of promoting clean energy innovation and deployment. The CEM programs, such as the SEAD initiative and the Clean Energy Solutions Center, have catalyzed the adoption of energy-efficient technologies and have facilitated policy development.

The International Energy Charter: Adopted in 1991, the International Energy Charter is a multilateral initiative that furthers energy cooperation among countries and organizations. The Charter has focused on energy transit, investment protection, and trade, thus helping facilitate dialogue and collaboration on energy security and other related matters. The ECT underpins the legal framework for energy trade and investment, thereby contributing to stability and predictability in international energy markets.

The Global CCS Institute: Of late, the Global Carbon Capture and Storage CCS Institute stands out as the epitome of explicit attempts by international actors to handle global climate change and energy security. The Institute has, since its founding year in 2009, remained at the forefront in promoting and facilitating the development and deployment of CCS technologies that will reduce these gas emissions. Through research, advocacy, and partnerships, the Institute has facilitated the advancement of the projects on CCS; thus, enhancing the global efforts to address climate change.

Recommendations to Reinforce the Role of International Institutions

To develop international institutions' role in energy security, one could proceed as follows:

Coordination and collaboration: No doubt, the coordination and collaboration of international institutions will multiply their effectiveness. It may be done by establishing formal arrangements through which information is shared, joint activities are undertaken, and integrated programs are designed. Further, it can avoid wasteful duplication of effort and channel time and resources toward maximum impact. For example, on selected energy security issues, joint task forces or working groups can be established for close collaboration or to regularize the work.

Expand Engagement with Emerging Economies: International institutions will become more relevant and influential when engagement with emerging economies is made, challenges and opportunities of which are very different than OECD members. Support and initiatives targeted to needs would allow more inclusive and effective solutions for energy security. As an example, technical assistance, capacity building, and funding support for the emerging economies may help to transition toward a cleaner energy system and enhance energy security.

Encourage Transparency and Precision of Data: Transparency and accuracy in reporting should be encouraged to ensure confidence and credibility. International institutions must put in place appropriate mechanisms with regard to the collection of data, verification, and dissemination. The development of standardized methodologies and frameworks in reporting data will further enhance the reliability of global energy statistics by promoting better decision-making.

Foster Public-Private Partnerships: Public-private partnerships can go a long way in fostering energy security initiatives and finding global solutions to address global challenges. Cooperate with Industry Stakeholders

Successes and Impact of International Institutions

International institutions have contributed a lot to energy security in one form or the other, in terms of functions or activities.

Energy Cooperation: This area is one of the most important areas in international institutions, which have been very successful in fostering cooperation among countries. For example, through the IEA energy dialogues and IRENA renewable energy forums, countries cooperate in areas such as energy policies, exchange best practices in governance on common issues such as infrastructure vulnerabilities, among others. For example, the IEA Energy Technology Perspectives report issued annually covers the trends of innovative technologies and policy developments; hence, allowing the member countries to understand which policy to implement and in what area to invest in innovative solutions.

Geopolitical tensions: This can work inversely to the effectiveness of international institutions; as geopolitical tensions arise from conflicting national interests. Countries may place their individual energy security imperatives over and above the collective goals and could therefore emerge with their positions in discord, making consensus elude easily. For example, differences in energy policy and energy interest among IEA member countries sometimes result in difficulties in devising common strategies and responses.

Limited Resources: Most of these international organizations have limited resources and budgets, which influence the implementation of the various mandates. With their minimum budgets come minimal activities, an inability to conduct exhaustive research, and a reduced potential to assist member countries. For example, perhaps IRENA is under budgeted to the point that it cannot extensively support and provide technical assistance to all its member countries, particularly those that have the least amount of resources.

Data Accuracy and Transparency: One critical challenge that has often faced international institutions is the accuracy and transparency of the data provided. Partial or inaccurate data can easily undermine the credibility of their analyses and recommendations. For example, differences in the way different countries report energy production and consumption affect the accuracy of world energy statistics, hence influencing policy decisions in the sector.

Coordinating and Integrating: It may be a challenge to how several international institutions coordinate their efforts and integrate their activities. Consequently, overlapping mandates, divergent priorities, and miscommunication among the institutions result in failures and redundancies. As an example, renewable energy initiatives of IRENA and the WEC need coordination towards the harmonization of goals and activities to avoid redundancy and impact more considerably.

CHAPTER TWELVE

The Role of Emerging Economies in Shaping Global Energy Security

"Energy is an issue that connects with poverty, hunger, education and equality"

Jim Yong Kim

The emerging world thus stands at the prime position in shaping global energy markets, policies, and security dynamics. As the economies of these nations continue to grow and expand, so does their energy consumption, thus making them critical stakeholders in shaping a secure future of energy globally. This chapter reviews the role of emerging economies in shaping global energy security and discusses emerging challenges and opportunities for the future of energy governance.

Increasing Influence of Emerging Economies

Economic Growth and Energy Demand: Large-scale emerging economies in the regions of Asia, Latin America, and Africa are seeing rapid economic growth, and, as a result, sharp increases in energy demand. Large populations, coupled with growing industrial sectors, make countries like China and India among the largest energy-consuming nations in the world. For example, China is the world's most consuming nation, with an increase in energy demand hand in hand with its economic growth. In contrast, the growing economy and increasing population have raised the energy requirements of India to become a leading player in world energy markets.

Strategic Energy Investments: Emerging economies are also making strategic investments in energy infrastructure and resources. These include the development of new energy projects, the acquisition of energy assets abroad, and increasing domestic energy production. For instance, China has gone big time into overseas energy assets through its so-called BRI aiming at the assurance of energy supplies to support infrastructure development globally. Such investments are very vital in diversifying energy supplies in the emerging economy and reduce dependence on any particular provider or region.

Impact of Energy Prices and Markets: With continued growth in most emerging economies, they tend to have strong impacts on global energy prices and market dynamics. Their energy consumption, in turn growing, might affect global energy prices, disruptions in supply chains, and economic instability around the world. For instance, emerging economies are increasing demand for energy, boosting oil and gas prices, shaping global markets and even the balance of energy trade. Conversely, these economies themselves may influence the price of energy through their own policies, trade agreements, and investment decisions.

Infrastructure Development: Most of the emerging economies face the gigantic challenge of developing and modernizing the energy infrastructure. This has the potential to cause inefficiencies, disrupting supply and creating difficulties in meeting the rising demand for energy. For example, countries with weak electricity grids or very limited access to sources of energy have tended to struggle to provide their people with adequate and affordable energy. Therefore, investment in

Energy Transition and Sustainability: Several emerging economies today, while growing economically, are trying to balance economic development with environmental sustainability. Such transition to cleaner energy sources therefore opens up a spate of opportunities as well as challenges simultaneously. While renewable energy does offer an avenue to reduce the carbon footprint and thereby mitigate environmental degradation, high upfront costs and limited availability of technology, along with insufficient regulatory mechanisms, stand as barriers toward the attainment of this end. In fact, all these transitions-whether to renewable energy from coal in highly coal-intensive countries, for example-are complicated and require huge investments in process and policy.

In addition, geopolitical and economic vulnerabilities are emerging in a number of emerging economies due to political instability, regional conflicts, or economic fluctuations that would pose an impact on energy security and result in uncertainties within energy markets. For example, countries featuring political instability or with ongoing conflicts may encounter disruptions in energy supplies and challenges in attracting foreign investment. Consequently, addressing such vulnerabilities will require strategic planning, international cooperation, and resilience-building measures.

Access to Technology and Innovation: Inadequate access to modern energy technologies and associated innovations are major stumbling blocks to the development and progress of energy systems in emerging economies. Technological change could act as a driver for increasing efficiency and sustainability in energy. However, it is increasingly difficult for emerging economies to adopt and assimilate new technologies as costs have risen and also require considerable technical expertise and infrastructure. Technology gap: international collaboration and technology transfer help in the development of more advanced and sustainable energy systems.

Opportunities for Emerging Economies: Leadership in Renewable Energies. The economies of the emerging markets can emerge as leaders in renewable energies at a global level. There are countries endowed with a good share of natural resources, such as solar power, wind, and hydroelectricity. These resources can be used by countries to reach carbon neutral while meeting their energy demand. For example, the Indian government and the Brazilian government have invested hugely in solar and wind power projects. They make them the leading countries in renewable energy generation. These projects contribute not only to saving the world from degradation but also toward economic development and further energy security.

Regional Energy Cooperation: Regional cooperation can, indeed, enhance energy security and support sustainable development in emerging economies. Collaboration on a number of initiatives, such as regional power pools, cross-border energy projects, and shared infrastructure, will meet the common challenge and make better use of resources. For instance, the Programme for Infrastructure Development in Africa of the African Union has a plan for promoting regional energy connectivity and integration that would enhance economic development and energy security throughout the continent.

Leapfrogging and Technological Innovation: This would offer the opportunity for leapfrogging indigenous energy technologies and realizing innovative solutions that meet contextual needs. In countries lacking widespread access to electricity, decentralized energy systems can be deployed-solar home systems and mini-grids-in order to sell or provide services with this power to under-served communities. With these new technologies and business models, emerging economies are in a position to accelerate energy transitions and achieve relevant sustainable development goals.

Investment and Economic Growth: The emerging economies can attract investment in energy sectors to achieve economic growth and development. Foreign direct investment in the energy projects provides much-needed capital, technology, and expertise to support infrastructural development and resource management. In this regard, investment in oil and gas exploration, renewable projects, and energy efficiency can create jobs, stimulate economic activities, and improve energy security.

Case Studies of Emerging Economies Shaping Energy Security

China's Energy Strategy: As newer influential emerging economies go; China's energy strategy epitomizes the new influence which they are getting to wield in global energy security. Large-scale initiatives by China, such as the Belt and Road Initiative, and investments in clean energy technologies gradually shape the global energy markets and security dynamics. The BRI, on one hand, aims at supply security across a number of regions, coupled with infrastructure connectivity, while China's investments in renewable energy projects, on the other hand, contribute to global sustainability efforts. This is a pointer to how emerging economies can use their economic heft to influence global energy trends and policies.

India's Renewable Energy Expansion: India typifies how emerging economies can take center stage in sustainable development. The country has declared ambitious targets to expand its renewable energy capacity in solar and wind power projects. Policy frameworks, such as the National Solar Mission and the National Wind Energy Mission, have therefore been in place to provide the necessary support to India's commitments toward renewable energy. This, in turn, enables increased deployment of clean energy and helps bring down carbon emissions. India's efforts underline how far the emerging economies can go in driving global energy transitions and contributing toward energy security.

Brazil's Bioenergy Sector: Bioenergy development in Brazil considers emerging economies vital drivers for sustainable energy solutions. Its biofuels industry, mainly ethanol derived from sugarcane, has positioned Brazil to be leading with respect to renewable energy. The Brazilian experience with the production of bioenergy reveals that emerging economies can transform their indigenous resources and new technologies into opportunities to enhance their energy security while reducing dependence on fossil fuels.

South Africa Energy Mix: Indeed, the South African case epitomizes the range of experience in energy mix diversification across the emerging economies. Moreover, it is heeding a call to balance its overdependence on coal with investment in renewable energy and natural gas. Meanwhile, it provides that the IRP offers a strategic vision for South Africa toward a more sustainable energy system: attaining energy access, reliability, and environmental concerns. The South African experience epitomizes the challenge of managing energy transitions and has required strategic planning and policy support.

Strategic Recommendations for Emerging Economies

The new emerging economies need to formulate comprehensive policies on energy issues, keeping in view the aspect of energy security, sustainable development, and economic growth. Comprehensive policies would require well-defined goals, regimes, and incentives in support of clean energy deployment, infrastructure development, and technology innovation. For instance, feed-in tariffs, renewable portfolio standards, and minimum energy efficiency standards can serve as effective catalysts for the development of clean energy sectors, thereby allowing for greater depth in energy security.

Strengthening regional cooperation will improve energy security and add to the sustainable development of emerging economies. This would facilitate regional cooperation among countries over regional energy projects that involve cross-border power transmission lines, common infrastructure, and joint initiatives at the regional level. For instance, regional power pools and integrated electricity grids can be established that would increase access to energy, reducing costs and thereby improving reliability.

Invest in Infrastructure Development: Energy infrastructure investments will go a long way in helping the attainment of economic growth with reliable energy supplies. It is important that infrastructure development be one of the priority areas of emerging economies, covering grid modernization to energy storage and transportation networks. Public-private partnership and international financing mechanisms can provide capital and expertise for such infrastructure projects.

Promotion of Technology Transfer and Innovation: Such roles of technology transfer and innovation will go a long way in helping overcome particular barriers in the development and deployment of clean energy and infrastructure. The sharing of knowledge, technologies, and best practices needs to be supported by governments and organizations. In this regard, collaboration with international organizations and technology providers could establish rapid deployment of advanced energy technologies or accelerate innovation.

Address Geopolitical and Economic Risks: Methods to manage geopolitical and economic risks impacting energy security need to be developed by emerging economies, diversifying the source of energy, building strategic reserves, and using diplomatic efforts to minimize conflicts and disruptions. The building of contingency plans and resilience measures will meet future challenges and needs as far as energy security is concerned arising out of uncertainties.

Improve Public Understanding and Awareness: Public awareness and understanding are essential in rallying people's support for energy policy and programs. Emerging economies need to inform the public about concerns on energy conservation, efficiency, and sustainability. Educational programs and public campaigns may eventually bring behavioral change, which would further support policy intentions and help ensure wider community involvement in energy transition.

Conclusion

Emerging economies are increasingly driving the global energy environment, from markets to policies to security dynamics. With continued economic growth and increasing energy use by these countries, they will play an important role in contributing to global energy security. By addressing challenges, capturing opportunities, and taking strategic action, emerging economies can make valued contributions toward a secure and sustainable energy future. Investment in infrastructure, improved technology, regional cooperation, and good policies are some of the ways emerging economies can try to work their way around the complications in the world's energy landscape to provide energy security to their people and the rest of the world.

CHAPTER THIRTEEN

THE PATH FORWARD: STRATEGIES FOR ENHANCING GLOBAL ENERGY SECURITY

"The future belongs to those who innovate, not to those who imitate"

Santosh Kalwar

When the world energy scene is in ever-changing evolution, energy security demands a multi-faceted approach dealing with technological, geopolitical, economic, and environmental problems. This chapter therefore evaluates strategic pathways for enhancing global energy security by focusing on innovative solutions, collaborative measures, and proactive steps toward solving the uncertainties of the future. Building on contemporary trends, pin-pointing opportunities, and putting forward certain concrete proposals, the chapter tries to give a comprehensive framework for energy security strengthening at the global plane. Improve Energy Infrastructure: Modernization of Energy Infrastructure is one of the most significant strategies for boosting energy security through updating and upscaling energy infrastructure.

Aging infrastructure can be prone to disruptions and inefficiencies, affecting the reliability and stability of the supplies. Major infrastructure improvements will involve investments in smart grids, efficient transmission systems, and resilient power generation facilities. For example, further integration of more digital technologies into grid management could significantly enhance real-time monitoring, fault detection, and response capabilities, thereby reducing the likelihood of outages and improving system reliability as a whole. Enhancing Resilience to Climate Change: The great risk of climate change involves rising temperatures, a higher frequency of extreme weather events, and rising sea levels, all of which fall upon energy infrastructure.

In building climate-resilient infrastructure, the name of the game is designing and constructing manifold facilities to resist environmental stressors. Power plants and substations can be sited on higher ground to decrease the vulnerability of their structures and operations to flooding, while energy storage can be sited at key locations to provide standby power during extreme weather. Integration of infrastructure planning and development with climate resilience will ensure uninterrupted supplies of energy in light of changed environmental considerations. Energy Source Diversification: One of the key strategies for diversification seeks to decrease dependence on a single type of fuel and improve energy security.

A diversified supply can also reduce vulnerability to supply disruption and price volatility. For instance, adding solar, wind, and hydroelectric power can increase the reliability of the electricity supply and provide a more stable source of electricity, while also reducing greenhouse gas emissions. Enhancing Technological Innovation

Research and Development: Investment in research and development is crucial because this maintains the tempo of development of energy technologies and further helps in ensuring energy security.

Innovations in energy storage, carbon capture, and renewable energy technologies could help resolve existing challenges and give way to new opportunities for security in supplies of energy. In addition, governments should cooperate with private sector organizations and research institutions in funding and supporting R&D projects to drive technology advancement. As one example, the development of various battery technologies could provide better energy storage capacity and reliability, which would mean greater inroads of renewable energy sources into the grid. Enabling Breakthrough Technologies: Hydrogen fuel cells, advanced nuclear reactors, and next-generation solar panels are technologies that can revolutionize the energy landscape of tomorrow. In enabling technologies lies a pathway to further security and sustainability in our energy futures. Hydrogen fuel cells, for instance, present a cleaner alternative to conventional fossil fuels for transportation, industrial processes, and the generation of power. This calls for collaboration by the governments and industrial stakeholders in the promotion of commercialization and adoption of emerging technologies. Improve Cybersecurity: With energy systems continued digitization and interconnectivity, cybersecurity has been deemed a critical element of energy security.

The energy infrastructures can be protected against the cyber threats through deployment of appropriate security measures, periodic assessment, and responding to emerging vulnerabilities. For example, advanced methods of encryption, intrusion detection systems, and regular security audits would go a long way in ensuring that critical infrastructures are protected from potential cyber-attacks. It will be important to develop appropriate strategies to address cybersecurity challenges with collaboration between governments, private sector organizations, and cybersecurity experts. Smoothening International Cooperation

Strengthening Multilateral Treaties: This would play an important role in encouraging international cooperation over energy security matters.

The Paris Agreement, the International Energy Charter, and regional frameworks of energy cooperation provide a platform for countries to come together toward mutual goals and address common challenges. These agreements further strengthen and develop new frameworks that may assist in bringing about global energy security by promoting coordinated actions, sharing best practices, and reinforcing collective efforts. In other words, the reinforcement level of commitments for greenhouse gas emissions reductions can also serve to further reinforce effectiveness at the Paris Agreement and help bring about a more stable and sustainable global energy system. Strengthening Regional Cooperation: Specific energy security challenges can be addressed through the approach of regional cooperation, which, in turn, can foster stability in geographic areas. Regional organizations, such as the Energy Union by the European Union, the African Energy Commission by the African Union, and the APEC, provide a distinct platform on which countries can cooperate on energy issues, share their resources, and bring harmonization among their policies. Mechanisms of regional cooperation must be consolidated and, wherever felt necessary, should be extended to respond to cross-border energy challenges, infrastructural connectivity, and hence energy security issues related to the regions.

Public-Private Partnerships: PPPs can provide a lot in furthering energy security through the sharing of resources, expertise, and innovation between the two sectors. Such projects involving governments alone or combined with private companies and NGOs will lead to the development and deployment of new technologies, modernization of infrastructure, and development of policy and regulatory frameworks. As part of renewable energy installation, for example, PPPs could facilitate large-scale solar and wind projects. They can also encourage research into the development of new technologies.

Implementing Effective Policies and Regulations

Development of Full-Scale Energy Policy: "In developing full-scale energy policy, effective energy policy helps to steer the transition toward a more secure and sustainable energy system.

The policymakers should articulate an all-inclusive approach that gives due consideration to energy security, environmental sustainability, and economic development. Effective policy can be secured by clearly defining the goals, establishing a regulatory framework, and offering incentives for investment in clean energy technologies. Policies on energy efficiency standards, renewable portfolio standards, and carbon pricing could drive the use of sustainable options and improve energy security. Energy Efficiency Promotion: Energy efficiency can help to drastically reduce energy consumption and cost, therefore such a measure would assure energy security. Application of the minimum standard of energy efficiency for buildings, appliances, and industrial processes can save much energy and reduce demand on energy infrastructures. Application of strict standards of efficiency for lighting and heating systems could reduce energy use and lower greenhouse gas emissions. The use of regulations, incentives, and public awareness by the government or organizations should promote energy efficiency.

Encouraging Energy Innovation: In order to rise to new challenges and further enhance energy security, innovation in energy technologies and practices has become an issue of utmost importance. In light of this, it is relevant for governments and industry stakeholders to initiate support for the creation and diffusion process through funding programs, tax incentives, and research grants. Financial incentives could be made available to pilot projects and demonstration initiatives as a means to accelerate the commercialization process and start wider dissemination.

Addressing Geopolitical and Economic Challenges

Mitigating Geopolitical Risks: Geopolitical risks in the form of political turmoil, disputes over territory, tensions in trade relations, among others, may be transmitted to the energy function by supply disruptions and changes in energy prices.

In this regard, countries should try to minimize these risks through diplomatic engagement, enhanced international partnerships, and contingency planning. For example, vulnerability to geopolitical disruption would be lessened by diversification in importation sources of energy and investment in strategic reserves. Furthermore, the facilitation of dialogue and cooperation with major energy-producing and consuming countries would enhance stability in world energy markets, as it would enhance the prospect of resolving potential conflicts. Ways of managing energy price volatility are varied; it could affect economic stability and energy security through production cost and consumer price fluctuations. The government and businesses have to engage in ways to reduce or manage price volatility, which includes the establishment of price stabilization mechanisms, investment in energy storage, and diversification into different energy sources. Examples are developing strategic petroleum reserves and long-term contracts with supplies of energy to provide stable oil prices to protect the people and businesses from market disruption impacts. Promotion of economic diversification: in general, the diversification of economies can raise levels of energy security with the lessened reliance on a single energy source or energy market.

In an economy, diversification allows countries to be resilient concerning the determination of energy prices, disruptions of supply, and volatility of the market. This could be realized through, among others, investment in other sectors like technology, manufacturing, and agriculture to create other streams of revenue independent of energy export. Economic diversification will lead to resilience in light of these challenges associated with energy. Stakeholder and Public Engagement

Engaging Stakeholders in Decision-Making: The effective engagement of stakeholders-industry leaders, community groups, and environmental organizations-is of utmost importance in the formulation of energy security strategies. Engaging all these stakeholders in the decision-making process can achieve a range of perspectives and concerns, building consensus on energy policy and initiatives. There can be significant improvements in results from engaging local communities in planning and developing energy projects considering better outcomes of the projects and the support of the public for the same. Engagement with stakeholders could lead to increased transparency and accountability in decision-making within the energy sector.

Public Awareness and Education: Public awareness and education are invaluable in the drive towards energy security and good environmental stewardship. This can reinforce policy goals by increasing awareness on energy issues, especially on conservation and efficiency, renewable energy, and behavioral changes. Public campaigns on the benefits of energy efficiency measures and the adoption of renewable energies enable individuals and businesses to take proactive steps toward reducing energy consumption and increasing support for clean energy initiatives.

This means developing a competent workforce who will help ensure that energy security and transition towards sustainable energy system will keep on being advanced.

It is also important to invest in education and training programs for energy professionals, which will address the competence gaps. This will enable technological innovation and support the industrial capacity, such as training technicians, engineers, and researchers to attain skills and know-how in emerging energy technologies, enabling the growth of clean energy industries. Such workforce development initiatives might further spur economic growth and increase the capacity for the nation to address energy security challenges. Conclusion

Improving energy security is multi-dimensional: technological, geopolitical, economic, and environmental challenges all need to be overcome.

It means modernization of energy infrastructures, technological innovation, international cooperation, efficient policies and measures to mitigate geopolitical and economic risks, and interaction with stakeholders that, in turn, will enhance the energy security of nations and organizations toward a more resilient and sustainable energy future. The way forward requires striking a balance between diverse priorities, leveraging collaborative effort, and adapting to continuously changing trends and challenges. In sum, the paths to solving such complexities in the global energy landscape for a secure and sustainable supply for future generations involve strategic planning, innovative solutions, and necessary proactive measures.

CHAPTER FOURTEEN

Navigating Energy Security in a Changing Climate: Adaptation and Resilience Strategies

"Hope for the best but prepare for the worst"

Benjamin Disraeli

Climate change affects global energy security on a wide range of topics, from availability and reliability of energy resources to the overall economic and environmental context in which energy systems function. In the face of the inevitable changes in the climate, it is increasingly important to propose and implement strategies that increase resilience and adaptability against changing risks. This chapter discusses the nexus of climate change and energy security, considering approaches to adaptation and resilience which may provide routes to stability and sustainability within the energy systems.

Understanding the Impacts of Climate Change on Energy Security

Climate change gives rise to physical risks in energy infrastructure, increased frequency, and intensity in extreme events such as hurricanes, floods, and heatwaves. The consequences are damage to energy facilities, disruption of supply chains, and also power outages. Of the named ones, hurricanes may badly jolt oil refineries and power plants, thus disrupting the supply of energy on a large scale. Understanding these physical risks is necessary for mitigation in order to maintain resilience in energy infrastructure.

Changes in Energy Demand: Climate change can alter the patterns of energy demand that, in turn, alter heating and cooling needs, water availability for hydropower, and seasonal energy consumption. Rising temperatures may raise demand for cooling, while altered precipitation patterns will impact the availability of water for hydroelectric generation. For instance, prolonged drought may lower the water level in reservoirs, which in turn impairs the normal operation of hydropower plants. All of these changes also require energy systems that are flexible and responsive to shifting demand patterns.

Energy Resource Availability: There are several ways in which energy resources may be affected by climate change, including those related to fossil fuels, renewable sources, and water resources. Changes in temperature and precipitation along with rising sea levels can influence the extraction, production, and transport of energy resources. For instance, the melting of Arctic ice may open new shipping routes for oil and gas, yet at the same time, environmental and geopolitical risks abound. Sustainability in the management of energy resources in light of climate change calls for comprehensive planning and assessments of risks.

Strategies for Adapting Energy Infrastructure Resilience of Infrastructure: In order to adapt to climate change, energy infrastructure has to be resilient and strong enough in the face of impacts caused by it. This involves the consideration of climate at the planning, design, and construction phases of energy facilities. Power plants and transmission lines can be raised in flood-prone areas, while those that involve heat can adopt adaptive cooling mechanisms to deal with high temperatures. Building climate-resilient infrastructure also requires the use of appropriate materials and technologies to adapt infrastructure to extreme weather conditions while minimizing vulnerability.

Upgrading Grid Systems: Modernizing and upgrading electricity grids can improve their resilience to climate-related disruptions. Smart grid technologies, which would allow real-time monitoring and control of the grid, would afford greater agility in responding to and recovering from outages. Energy storage systems included in the upgrade could provide backup power during emergencies, apart from managing fluctuations in renewable energy generation. Strengthening the grid infrastructure and introducing contemporary technologies can ultimately increase the overall reliability and flexibility of energy systems.

Diversification into Energy Sources: It would reduce overdependence on a particular source and hence make it resilient to disruption caused by climate. The incorporation of a mix between renewable energy, natural gas, nuclear power, and energy storage would create a more constant and reliable energy supply. This can be seen by the use of solar and wind in conjunction with battery storage, which offers a better potential to balance out the intermittence of generation from renewable sources, hence providing a stable supply of energy. Energy source diversification also includes the development of new technologies and alternative resources that could deal with potential risk and uncertainty.

Increasing Resilience Through Policy and Planning

Climate-Resilient Policy Development: It gives the impetus to the policy framer with a view to developing a climate-resilient energy policy. The policy will address issues on climate risks, adaptation measures, and low-carbon transition of energy. Examples include policies that incentivize the adoption of energy-efficient technologies, promote the deployment of renewable energies, and foster adaptation measures to enhance general energy security due to climate change. Efficient policy frameworks should also allow monitoring, evaluation, and updating of strategies based on newly emerging risks and developments.

Climate Risks Integrated into Energy Planning: Integration of assessment of climate-related risks into energy planning helps bring on board the vulnerabilities that need formulation of appropriate adaptation strategies. This involves analyzing the possible impacts of climate on the energy system, assessment of risks, and development of contingency plans. Such risk analysis in energy planning includes studying the potential impacts of rising sea levels on coastal infrastructure, assessment of the risks that drought poses to supplies dependent on water, and strategy formulation to reduce these risks. Integration of climate-related risks into planning makes energy systems well prepared for future challenges.

The enhancement of resilience through international cooperation is another sphere of action. Climate change is a global problem, and international cooperation is not an option but a task that must become reality to tackle its serious effects. Collaboration among countries, organizations, and institutions nurtures resilience, which supports adaptation measures. For example, different international agreements and frameworks, such as the Paris Agreement, provide various avenues whereby countries share knowledge and coordinate actions to support each other in responding to climate risks. This includes the sharing of best practices, technical assistance, and forming partnerships to enhance Global Energy Security.

Innovating Climate Resilience and Adaptation

Improvement of Renewable Energy Technologies: Inventions within renewable energy technologies can enable climate adaptation and build resilience in delivering more efficient and reliable sources of energy. Increased efficiencies in solar panel efficiency and better wind turbine design together with storage systems may be able to enhance the actual performance and reliability of projects in renewable energy. Investment in research and development regarding new technologies can enable transition to a more sustainable energy system and improve resilience to climate-related disruptions.

Climate-Resilient Energy Storage Solutions: Energy storage technologies have become very important to improve resilience by allowing backup power and balancing intermittent renewable generation. Hence, energy storage innovation, including advanced batteries, pumped hydro, and thermal energy storage, will go a long way in improving the reliability and flexibility of the energy system. The development of fully climate-resilient storage solutions faces scalability, cost, and performance challenges to be surmounted if the storage systems are to sustain energy security amidst an unpredictable changing climate.

Integration of Smart Grid and Digital Technologies Energy systems will be resilient through real-time monitoring, control, and optimization. Smart grids can provide critical data on energy consumption, grid performance, and system health that enables proactive management and response against disruptions. Sensors, analytics, and artificial intelligence are some examples of digital technologies that can be integrated into energy systems to improve their effectiveness and reliability through predictive maintenance, fault detection, and automated decision-making.

Climate Resilience in Building and Infrastructure Design: The integration of climate resilience in the design of building and infrastructure can reduce vulnerabilities to generally improve energy security. This covers material and construction techniques that are resilient in extreme weather conditions, efficiently designed building methods, and infrastructure designed with consideration for climate risks. Examples include adding green roofs, energy-efficient windows, and resilient HVAC systems to enhance the climate resilience of buildings and reduce their burden on energy demand.

Case Studies of Successful Adaptation and Resilience Strategies

Flood Management in the Netherlands: The Netherlands represents one of the best adaptations to climatic risk. The country invested in sea walls, levees, and storm surge barriers with an ascertained protection against sea level rise and extreme events. Community engagement, proactive planning, and innovative engineering are emphasized by the Dutch approach as necessary measures for ensuring energy infrastructure resilience and overall safety.

practices. The state's approach showcases the integration of risk assessments, technology advancements, and community engagement in building resilience to climate-related hazards.

Singapore's Integrated Energy and Climate Planning: Singapore's integrated energy and climate planning offers a case study on how a small nation can go about planning integrated energy and climate policies. California's Wildfire Resilience Measures: Various initiatives have been taken in California to establish better resilience to risks arising from wildfires affecting energy infrastructure and general energy security. It includes strengthening the utility infrastructure, developing advanced fire detection technologies, and managing vegetation to address climate risks with a view to further enhancing resilience. Very comprehensive strategies have been developed for management of energy resources, emission reduction, and adaptation to climate change. Its initiatives include investments in clean energy technologies, enhancing energy efficiency, and implementing climate-resilient infrastructure. This underlines the essence of holistic planning and proactive measures that need to be taken to achieve energy security and sustainability.

Energiewende of Germany: The German Energiewende is one of the ambitious attempts toward an energy system with low-carbon transition and addressing various impacts of climate change. The governments have been working on expanding renewable energy, efficiency in energy use, and modernizing their grid. Germany's experience very well illustrates how ambitious climate policies, technological innovation coupled with long-term planning, enhance energy security and build resilience.

Recommendations toward Building Climate Resilience in the Energy Sectors

Conduct Integrative Climate Risk Analyses: Indeed, regular and detailed climate risk analysis is needed to identify the root of potential vulnerabilities across energy systems. From extreme weather events to shifts in resource availability and further into the patterns of energy demand, these analyses need to range across a large, broad scope of climate impacts. Such an exercise will enable the stakeholders to identify specific risks and vulnerabilities that, in turn, would make the adaptation strategy more focused, with the intention of reducing potential disruptions and enhancing resilience.

Invest in Climate-Resilient Infrastructure: Infrastructure investments should be resilient to the changing climate. This includes modernization of energy facilities, reinforcing power grids, and consideration of extreme weather in design standards. The building of resilient infrastructure to climate change concerns the adoption of technology and construction methods that improve durability, reduce vulnerabilities, and enable operation during climate adversities.

Support Research and Development of Adaptive Technologies: In this regard, it is important that more investment in research and development be directed toward technologies that enhance climate resilience and adaptability. This includes furtherance of energy storage solutions, renewable energy technologies, and smart grid innovations. Supporting R&D will help develop novel tools and systems that enhance flexibility and reliability in energy infrastructure, making it far more capable of responding to climatic-related stressors.

Integrated Policies in Climate and Energy: Design and implement such policies that help integrate climate adaptation objectives with the objectives of energy security. These should provide momentum for transition towards cleaner energy resources, infrastructural upgrade, and incentives for climate-resilient technologies. Laying down energy policies against their climate counterparts will help governments and organizations ensure that their energy systems are resilient enough and adaptive to the shift in climate conditions.

Encourage Collaboration Across Sectors: Enable governmental agencies, private sector businesses, research institutions, and community organizations to collaborate in a regular process. This cross-sector collaboration can support the sharing of information and resources relevant to the development of best practices for climate resilience. In collaboration, these actors will be able to constructively address a wide array of interconnected problems and find creative solutions to help improve energy security.

Community Involvement and Public Awareness: There is a need for community involvement and public awareness of the essence of incorporating climate resilience in energy systems. Community involvement and public information campaigns can help in building political support for the adaptation measures, which promotes proactive behavior. A more educated and informed population is able to participate in resilience efforts, as well as to call for policies that enhance security in the energy sector.

Adaptive Management Strategies: Include adaptive management strategies that allow for flexibility and responsiveness in the face of changed conditions. This includes monitoring and evaluation of climate impacts, updating plans and policies with newfound information, and adjusting operational practices. In sum, an adaptive approach ensures continuous evolution of energy systems and their resiliency toward emerging climate-related risks.

Promote Resilient Supply Chains: Improve the resilience of energy supply chains through diversified sources, reduced dependencies on particular sources, and increased logistical capabilities. This shall involve securing a number of supplies, investment in alternative routes of transportation, and building contingency plans for disruptions to supplies. In this regard, a resilient supply chain ensures a stable, reliable supply of energy, even against climate-related disruptions.

Climate Resilience in Energy Planning: Embed consideration of climate resilience in the long-term energy planning processes. This involves assessing potential climate-related vulnerabilities of energy systems, determining adaptation objectives, and then strategizing on how to manage those vulnerabilities in the future. Integrating resiliency into planning allows one to acknowledge that energy systems need to be designed and operated with full awareness of climate uncertainties and challenges.

Support Policy Innovation and Advocacy: Pursue innovative policy measures and programs that contribute to climate resilience and energy security. It could involve the promotion of new regulations, mechanisms of funding, and incentive systems to trigger action for the adoption of adaptive technologies and practices. Support policy innovation will thereby create an enabling environment for the building up of resilience, ensuring that all resources and support needed are indeed made available to them.

Elaborate metrics and indicators of monitoring and reporting on the effectiveness of resilience measures. Tracking progress and impacts of adaptation strategies can provide valuable information on effectiveness and guide future improvements in the effectiveness of such measures. The regular reporting on resilience metrics is important for accountability and furthers continued work towards full energy security.

Leverage international collaboration and funding: pursue opportunities for international collaboration and funding for climate resilience initiatives. Engaging with global partners, participating in international forums, and accessing funding programs provide additional resources and expertise to the resilience-building efforts. International cooperation reinforces knowledge sharing, various common challenges, and creates new opportunities for enhancing global energy security.

Engage in Public Awareness and Involvement: Public awareness and involvement can be enlisted in a grassroots effort to help public attention shift to security. Balance is corrected because it helps advocate policies of competitive, reliable energy for all.

Following these recommendations can take stakeholders closer to the resilience of energy systems, lower vulnerability to climate change, and help achieve an assured and sustainable energy future.

CHAPTER FIFTEEN

THE PATH FORWARD: SHAPING A RESILIENT AND SUSTAINABLE ENERGY FUTURE

"The best way to predict the future is to create It"

PETER DRUCKER

The way forward in the future of energy security certainly needs to chart a course through a complex landscape of both challenges and opportunities. Climate change, geopolitics, technology, and socio-economic factors interact in a way that a strategic approach will have to be laid down for a resilient and sustainable energy future. This chapter examines the essential components of this direction, revealing ways in which stakeholders can continue to cooperate, invent, and guide toward a secure and sustainable energy system. A Broad-Based Perspective on Energy Security

This, therefore, means that integration of climate and energy goals into one cohesive strategy will be what the future of energy security is hinged on. To address urgency on both fronts, namely climate change and essential requirements for reliable and affordable energy, unity of approach is an important ingredient that needs to be brought to the forefront. Policies and actions must concurrently face greenhouse gas emissions, the promotion of renewable energy sources, and improvement of energy resilience. This can be realized through comprehensive frameworks that integrate national and global commitments to climate action into the imperatives of energy security.

Systems Thinking: A system thinking approach has to be fitted to understand the interdependencies and interactions within the energy landscape. Energy systems interact closely with other sectors such as transportation, agriculture, and industry. It is a holistic view that helps identify potential risks, opportunities, and synergies that are not seen otherwise if the individual components are looked upon in isolation. Systems thinking thus ensures better planning, decision-making, and policy development.

Enhancing Multilateral Cooperation: The global nature of energy security challenges requires the strengthening of multilateral cooperation. International agreements, partnerships, and joint initiatives have a very important role in finding solutions to the following common challenges: climate change, resource management, and energy access. In such a way, countries can put together their resources, knowledge, and competencies to elaborate coordinated responses against global energy security challenges.

Furthering Technology Innovation and Deployment

Accelerating the Deployment of Renewable Energy Technologies: Accelerating the deployment of renewable energy technologies holds the very core in creating a sustainable energy future. Invention in solar, wind, geothermal, and other forms can drive down costs, increase efficiency and improve access. Support by governments and businesses through favorable policy environments is required to facilitate research and development for the scaling up of renewable projects.

Enhancing Technological Innovation: Technological innovation has been important to address the evolving challenges of energy security. Indeed, investment in leading-edge technologies can further enhance efficiency, reliability, and sustainability for energy systems in the form of advanced energy storage, CCS, and smart grids. It is about the public and private sectors developing those technologies, removing barriers from the diffusion process, and thus accelerating the transition to a low-carbon energy future even faster.

Digitalization and Smart Technologies: Digital technologies and smart systems can transform energy management by enhancing resilience. Smart grids, demand response systems, and advanced data analytics allow for real-time monitoring, optimization, and control of energy infrastructure. Indeed, digitalization can ensure an increase in grid stability, management of renewable variability, and higher energy efficiency. Strengthening Policy and Regulatory Frameworks

Full Implementation of Integrated Energy Policy: The road to transition to a secure and sustainable energy future effectively should be guided by appropriate energy policies. The policymakers should integrate an all-inclusive strategy that addresses energy security, climate protection, and economic development. The policies should henceforth encourage renewable energies, promote energy efficiency, and encourage innovation in technology. Long-term, stable, and transparent policy frameworks create certainty and attract investment in energy infrastructure and low-carbon technologies.

Encouraging Market-Based Solutions: Many market-based solutions, including carbon pricing and emissions trading systems, can be combined with renewable energy incentives to serve as drivers of positive changes in the energy system. These mechanisms create economic incentives for reducing emissions, promoting clean energy, and improving energy efficiency. By aligning market signals with environmental and security goals, policymakers can support moving toward a more sustainable energy system.

Strengthening International Agreements: International agreements provide a vital force in framing global energy security and climate policies. It is of essence that commitment be further addressed in agreements like the Paris Agreement and SDGs for the coordination of efforts globally, as well as at individual country levels with common objectives. Such international cooperation and strengthened implementation will improve collective efforts toward securing a sustainable energy future.Promoting Resilience and Adaptation

Building Resilience in Energy Infrastructure:

Resilience is an increasingly crucial factor in energy security, given shifts in climate and extreme weather conditions. Investments into resilient infrastructure include those in climate-proofed energy facilities, upgraded grids, and diversified sources of energy supplies, thus reducing vulnerability. Resilience capacity building through infrastructure planning, design, and maintenance will surely increase tolerance to cope with disruptions or recover from them.

Resilience to Change: Energy systems should be dynamic concerning changing environmental, economic, and geopolitical conditions. An energy system will be able to make use of adaptive management approaches, policy frameworks, and responsive operational practices in order to deal with uncertainties and changing risks. As a matter of fact, it is the adaptive approaches that will ultimately ensure the energy system is relevant and strong in relation to scenarios of the future.

Community and local engagement would be meaningful for the building of resilience and would ensure that the energy system meets local needs. This is where communities should be integrated into energy planning, implementation, or decision-making to raise the resilience measures a notch higher and elicit greater community support for energy initiatives. Besides, it will help in establishing unique risks and opportunities to tune energy systems to specific regional and community contexts.

Equitable Access and Social Considerations

Addressing Energy Poverty: Equitable access to energy is key in the quest to achieve secure and sustainable energy. A household is said to be in a state of energy poverty when it lacks access to reliable and reasonable-priced energy. It requires an effective policy with programs that address energy-poverty targets, which are basically vulnerable populations. Efforts should now be directed toward the scaling up of access to clean and affordable energy services, promotion of energy efficiency, and provision of finance for income household sub-sectors.

Social Equity and Energy Transitions: Considering social equity is one of the critical variables in the transition toward a low-carbon energy system. Ensuring a fair distribution of the benefits arising out of energy transitions-ensuring that disadvantaged communities do not take a disproportionate impact of such transitions-is highly important to attain inclusiveness in just outcomes. In this respect, policymakers and other stakeholders should make considerations toward social equity in the planning of energy, investing in energy, and developing policies so that all kinds of communities benefit from advances in clean energy.

Engaging Diverse Stakeholders: In the case of energy security, social issues must be weighed with the participation of varied stakeholders who comprise marginalized communities, indigenous groups, and NGOs. It ensures that a wide array of perspectives and needs are taken into consideration in decision-making processes. Such diverse voices can come together in collaborative approaches that will eventually foster effective and fair solution building in the sphere of energy.

Success Stories: Cases of Strategies and Innovations that Work

The Scandinavian Model of Energy Security: Sweden, Norway, and Denmark also serve as examples of successful policies in integrated energy security. These countries have significantly invested in renewable energy, advanced grid systems, and energy efficiency measures. Emphasizing sustainable development, climate resilience, and innovative technologies, these nations have emerged as leading nations in energy security and sustainability.

Singapore Solution for Urban Energy: The solution for urban energy in Singapore highly represents the integration of energy security with climate adaptation in a highly compact city-state. Among the different solutions being innovatively implemented within the country's premises are smart grids, green building standards, and sustainable transportation systems. Strategies adopted by Singapore show ways cities can solve energy and climate problems using technology, policy, and planning.

California's Clean Energy Transition: The transition of California to a clean energy economy is an instructive example in the quest for successful energy policy and innovation. It has set ambitious renewable energy targets, adopted carbon pricing mechanisms, and provided significant investment in clean technologies. California's experience underlines the potential of bold policy initiatives and technological development to attain energy security with sustainability.

Germany's Energy Transition (Energiewende): The Energiewende of Germany forms an extended case study in the transition toward a low-carbon energy system. The country has pursued aggressive targets for renewable energy, taken up energy efficiency measures, and supported technological innovation. The German experience offers a number of lessons on how long-term planning, policy coherence, and stakeholder engagement can pay off in successfully attaining a secure and sustainable energy future.

The Way Forward: A Call to Action

In this future-a future dictated by the dual imperatives of energy security and climate resilience-urgent action is called for from all quarters. Though formidable, the challenges ahead thus simultaneously open up opportunities to fashion a truly sustainable and secure energy future within our grasp. Here's a focused call to action for navigating this critical transition:

Set and Achieve Ambitious Goals: It is in reaching ambitious goals by stakeholders that the related and complex challenges of energy security and climate change can be effectively addressed. It is important that clear and quantifiable targets be set by governments, businesses, and organizations concerning reducing emissions, deploying renewable energy, and using energy efficiently. They would require actionable plans with committed resources so meaningful progress and transformation will be warranted.

Foster multi-stakeholder collaboration: Energy security and climate challenges are complex issues that require collaboration across sectors and borders. The governments, industry leaders, scientists, and civil society organizations need to join hands with each other to share not only knowledge but also to pool resources and coordinate activities in order to make sure that these issues are resolved. Strong collaboration might accelerate the pace of innovation, make the processes of policy implementation smooth, and raise voices for resilience-building processes.

It will drive technological innovation: every sustainable energy future is based on technological innovation. This investment in research and development is partly for emerging technologies associated with advanced energy storage systems, renewable energy, and smart grid solutions. Innovation will ensure a sustainable, efficient, and reliable energy system as stakeholders support it and facilitate the adoption of new technologies.

Implement Comprehensive Policies: Coherent and robust policies are needed to guide the transitions toward a secure and sustainable energy future. There is a need for policymakers to come up with and implement coherent energy and climate policies that integrate resilience and sustainability goals. These policies should include policies for promoting clean energy, making infrastructure more resilient, and facilitating adaptation strategies.

Engage and Empower Communities: Effective energy solutions will have to reflect the needs and aspirations of the local communities. For this, engaging communities in decision-making processes and empowering them to participate in energy planning and implementation could result in more inclusive and effective outcomes. Moreover, community involvement in the development helps in balancing social equity in terms of how not only one segment is heard.

Awareness and education of the masses on various important issues regarding energy security and climate resilience will also help build support for and stimulate appropriate action. The educational activities should emphasize state-of-art sustainable energy practices that make sense, the benefits from resilience measures, and how citizens and organizations can play a role. A knowledgeable and engaged citizenry is sure to understand, support, and advocate for the changes that must be carried out.

Monitor progress and adjust strategies: Strategies and measures must be under continuous monitoring and evaluation regarding their effectiveness. It is important that the stakeholders periodically assess progress attained, challenges arising, and adjust approaches duly. In this respect, the stakeholders shall be in a position to respond flexibly to any new information and shifting conditions affecting the implementation process for keeping the energy systems resilient as well as adaptable.

We must commit to this path forward with collaboration and innovation. We will be part of our times by handling present-day difficulties and building a resilient and sustainable energy future for generations in store. This path forward signifies obligation and opportunity alike: obligation for there is no other way, and opportunity to engage in effort and common vision toward a better world.

GLOSSARY

Energy Security The secured adequacy of energy resources at acceptable prices without significant interruption to the economy and the socio-political structure. Energy Crisis A severe reduction in supply of energy resources, such as oil, gas, or electricity that creates shortages and surges in the price levels, threatening economic and social disruption. Renewable Energy

Energy from natural processes that are constantly replenished, such as solar, wind, hydro, and geothermal, providing a renewable alternative to fossil fuels.

Fossil Fuels

Natural resources, such as coal, oil, and natural gas, that were formed from the remains of ancient plants and animals. Fossil fuels are non-renewable and the primary contributors to greenhouse gases.

Greenhouse Gases (GHGs)

Gases responsible for global warming and climate change due to their heat-trapping abilities in the atmosphere. The main GHGs are CO2, CH4, and N2O.

Climate Change

The slow process of variations in temperature and precipitation, along with other aspects of the Earth's atmosphere, occurring over a very long period of time; primarily caused by human actions such as the combustion of fossil fuels and deforestation.

Supply Chain

The entire network of entities, resources, activities, and technologies involved in the production, transportation, and distribution of energy from its source to the end-consumer.

Geopolitics

The study of the effects of geography (human and physical) on international politics and relations, especially in regard to global energy markets and resources.

Energy Transition

Transition of the world energy system from fossil fuel-based systems to a mix of renewable and low-carbon energy sources. This is being done in order to minimize greenhouse gas emissions for climate change reduction

Energy Infrastructure

The representative systems, structures and other physical facilities used for energy production, distribution and consumption, such as power plants, pipelines and transmission lines.

Energy infrastructure and information systems protection against digital attacks that could disrupt the energy supply, cause economic damage, and endanger national security.

Energy Independence

The degree to which a country can supply itself with all of its energy needs without the need to depend on importation from other countries; it is usually associated with reduced vulnerability to global fluctuations in energy markets.

Energy Efficiency

Utilizing less energy for the production of a specific product or completion of a particular task, generally leading to reduced energy use and lesser environmental impacts.

Energy Diplomacy

The use of energy resources as a tool of international relations where countries negotiate with each other, forge partnerships, and at times alliances to secure energy supplies and have influence in global energy markets.

Critical Infrastructure

Those systems and assets, which are fundamental to the functioning of a society and economy, such as energy infrastructure, need protection against terrorism, natural disasters, and cyber-attacks among others.

Carbon Capture and Storage - CCS

Technology for capturing carbon dioxide from sources like power plants and industrial processes for storage underground so as not to release it into the atmosphere.

Smart Grids

Smart grid systems leverage state-of-the-art digital technologies to enable efficient monitoring and management of the flow of electricity, further increasing the share of renewable sources in the overall energy mix. Energy Poverty A condition whereby a household or community lacks reasonable access to adequate, affordable, reliable, and sustainable energy services so as to compromise quality of life and economic opportunities. Intermittency

Renewable energy sources, such as solar and wind, whose production varies with the prevailing environmental conditions. Strategic Reserves Storing reserves of energy resources, such as oil or natural gas, by governments to be used during emergencies or disruptions in supply for market stabilization and energy security. Paris Agreement

A global agreement under the UNFCCC that was achieved in 2015 which includes the aim to limit global warming to well below 2 Degrees above the pre-industrial levels.

Subsidies to energy

Government financial supports to the production or consumption processes of energy, which are many times given to reduce energy cost for the consumers or incentivize certain energy technologies.

REEs

A group of 17 chemically similar elements that are vital to the production of many high-technology devices and renewable energy technologies, including wind turbines and electric vehicles.

Chokepoints

Strategic narrow passages or routes - such as the Strait of Hormuz or the Suez Canal - through which global energy must flow. Disruptions at chokepoints have significant impacts on global energy supply.

Shale Revolution

The increase at an incredible rate in the United States of oil and natural gas production from shale formations due to the improvement in hydraulic fracturing-so-called fracking-and horizontal drilling technologies.

Just Transition

A framework concept for making a transition to the low-carbon economy equitably and inclusively by providing support for workers and communities impacted by the shift away from fossil fuels.

Hydraulic Fracturing (Fracking)

A drilling technique, in which high-pressure fluid is pumped into the rock, fracturing it and allowing the oil and natural gas to flow to the surface.

Energy Storage

Any number of technologies or systems serving as energy storage for subsequent use, which therefore can control the supply of the resource power generation, such as batteries, pumped hydro, and thermal storage. Base Load Power

The minimum quantity of demand on an electrical grid over a specific period. It typically comprises unbroken and dependable sources of power which the nuclear, coal, or natural gas plants can supply.Resilience

The competence of the energy systems to RESIST, absorb, and recover from different disturbances; natural disasters, cyber-attacks, and supply chain interruptions, amongst others, that guarantee the continuous supply of energy.

AGREEMENT AND TREATISE

The Energy Charter Treaty, ECT (1994)

Purpose The Energy Charter Treaty is a multilateral structure of energy cooperation that aims to facilitate energy security through open and competitive energy markets, protection of foreign investments within the energy sector, and dispute resolution mechanisms.

Significance: Legal stability and predictability of energy investment among member countries is very important for the ECT. The role of the ECT is also very crucial in facilitating international trade in energy and investment therein.

The Paris Agreement (2015)

Purpose: The Paris Agreement is an international agreement under the United Nations Framework Convention on Climate Change. It aims to contain global warming "well below 2 degrees Celsius above pre-industrial levels and pursuing efforts to limit it to 1.5 degrees".

Significance: Even as the focus of the Paris Agreement is on climate change, there is a great implication for energy security, as it encourages a world shift toward renewable energy and reduced dependency on fossil fuel energy, thus reshaping the globe's energy policy. Kyoto Protocol 1997

The Kyoto Protocol was an international agreement whereby nations that signed into it committed themselves to reduce the emission of greenhouse gases. This is out of scientific consensus that global warming is going on and human-made CO_2 drives it. Significance: It laid a foundation in integrating environmental consideration into energy security strategy, emphasizing the need to shift towards cleaner sources of energy.

IEA Treaty, 1974

The IEA had been formed in the aftermath of the 1973 oil crisis, and its charter was to maintain energy security for its member countries. The IEA treaty lays down measures that shall be taken by cooperation for stable energy supplies and contains emergency oil-sharing mechanisms and coordinated energy policies.

Significance: IEA played a vital role in monitoring world energy markets, proposing policies, and working together with its members on emergency response measures.

Treaty on the Functioning of the European Union -TFEU (Energy Chapter 2009)

Objective: The energy chapter of the TFEU is the legal foundation of the energy policy of the European Union. It focuses on the functioning of the energy market, energy efficiency, as well as security of supply.

Significance: This treaty is important to create an integrated energy market in the EU, reduce the dependence of the Union upon external sources of energy, and to further the transition toward renewable energy.

United Nations Convention on the Law of the Sea (UNCLOS) 1982

Purpose: UNCLOS sets the legal regime with regards to the use of the world's oceans and its resources. It also contains provisions regarding exploration and exploitation of energy resources found offshore, such as oil and gas, within EEZs.

Significance: UNCLOS plays an important role in terms of energy security, as it stipulates rights regarding exploration and exploitation of resources in the marine environment, with consequences for global energy supply routes and maritime security.

Statute of the Organization of the Petroleum Exporting Countries, 1961

The purposes of this Statute are to state the objectives and principles of the Organization of the Petroleum Exporting Countries and to provide the means by which these objectives and principles may be achieved, through the co-ordination of oil production policies amongst the member states with a view to stabilizing oil markets with a view to providing for the producers an income consistent with their requirements.

Significance: OPEC is the key determinant of global energy security since it affects the price level and level of oil production with high-impact consequences on the world energy markets.

Russia-Ukraine Gas Transit Agreement 2009 and 2019

Purpose: These agreements provide the legal framework within which the transiting of natural gas from Russia to Europe via Ukraine-a very crucial supply route to European energy security-is done.

Importance: These agreements are crucial in providing supplies of natural gas continuously to Europe with minimum disturbances that may be sparked by geopolitics.

Energy Initiatives of African Union Agenda 2063

Objective: Agenda 2063 is the African Union's strategic framework for the socio-economic transformation of Africa over 50 years. Energy initiatives are taken that need urgent action for improvement of energy access and enhancement of energy security with the promotion of sustainable energy development.

Significance: The energy programs under Agenda 2063 will help alleviate energy poverty in Africa through enhancing the level of independence in supplies of energy and expanding regional cooperation in energy.

Japan-Australia Economic Partnership Agreement (JAEPA) 2015

Purpose: JAEPA is a bilateral agreement in trade that involves cooperation in energy and resources primarily in the supply of resources of energy like LNG in a stable manner from Australia to Japan.

It shows the importance of bilateral partnership in pursuing energy security by reliable and long-term energy supply agreements.

REFERENCES

Chapter 1

- Azad, M. A. K., & Hossain, M. S. (2022). A review of energy security index dimensions and organization. *Energy Research Letters, 3*(1), 1-12. https://erl.scholasticahq.com/article/28914-a-review-of-energy-security-index-dimensions-and-organization
- Our World in Data. (2024). *Energy production and consumption.* https://ourworldindata.org/energy-production-consumption
- Ritchie, H. (2021). *How have the world's energy sources changed over the last two centuries? Our World in Data.* https://ourworldindata.org/global-energy-200-years
- Science Learning Hub. (2018). *Energy sources through time - timeline.* https://www.sciencelearn.org.nz/resources/1636-energy-sources-through-time-timeline
- World Economic Forum. (2022). *The 200-year history of mankind's energy transitions.* https://www.weforum.org/agenda/2022/04/visualizing-the-history-of-energy-transitions/

Chapter 2

- Cherp, A., & Jewell, J. (2019). *Conceptualizing energy security and resilience.* National Renewable Energy Laboratory (NREL). https://www.nrel.gov/docs/fy24osti/89206.pdf
- Khatib, H. (2010). *Competing dimensions of energy security: An international perspective.* Academia. https://www.academia.edu/40232908/Competing_Dimensions_of_Energy_Security_An_International_Perspective
- Tsiaras, A., & Gakis, D. (2019). Dimensions, components, and metrics of energy security: Review and synthesis.

SPOUDAI Journal of Economics and Business, 69(2-3),58-76. https://spoudai.unipi.gr/index.php/spoudai/article/viewFile/2755/2693

Chapter 3

- International Renewable Energy Agency. (2023). The geopolitics of the energy transition and opportunities for international cooperation. International RenewableEnergyAgency. https://www.irena.org/Publications/2024/Apr/Geopolitics-of-the-energy-transition-Energy-security
- Moran, D. (2011). Geopolitics and energy security: The inevitable connection. In Energy security and global politics: The militarization of resource management(pp.15-28).PalgraveMacmillan. https://doi.org/10.1057/9780230306851_2
- Moran, D. (2022). Geopolitics and energy security in Europe. In Energy security and global politics: The militarization of resource management (pp. 15-28). Palgrave Macmillan. https://doi.org/10.1057/9780230306851_2

Chapter 4

- Bagliani, M., & Marzo, P. (2020). Energy security and renewable energy: A geopolitical perspective. In Energy transition and energy security (pp. 1-22). IntechOpen. https://www.intechopen.com/chapters/71552
- International Energy Agency (IEA). (2020). Energy security in energy transitions. In World energy outlook 2022. https://www.iea.org/reports/world-energy-outlook-2022/energy-security-in-energy-transitions
- International Energy Forum (IEF) & Boston Consulting Group (BCG). (2016). Enhancing energy security: The role of technology. IEF Ministerial Roundtable. https://

- www.ief.org/_resources/files/events/ief15-ministerial/ief15-2nd-parallel-roundtable-ief-bcg-background-materials-27---28-september-2016-fin.pdf
- Wikipedia contributors. (n.d.). Energy security and renewable technology. Wikipedia,TheFreeEncyclopedia. https://en.wikipedia.org/wiki/Energy_security_and_renewable_technology

Chapter 5

- Atlantic Council. (n.d.). Geopolitics & energy security. Atlantic Council. https://www.atlanticcouncil.org/issue/geopolitics-energy-security/
- Deutch, J. (2008). The geopolitics of energy: From security to survival. Brookings Institution. https://www.brookings.edu/articles/the-geopolitics-of-energy-from-security-to-survival/
- International Renewable Energy Agency (IRENA). (2023). Geopolitics of the energytransition:Energysecurity.IRENA. https://www.irena.org/Publications/2024/Apr/Geopolitics-of-the-energy-transition-Energy-security
- Liu, W., & Lu, J. (2023). Geopolitics of the energy transition. Journal of Geographical Sciences, 33, 1-18. https://doi.org/10.1007/s11442-023-2101-2
- Mitchell, T., & Smith, A. (2023). The politics of energy security. Nature Energy, 8, 1096-1104. https://www.nature.com/articles/s41560-023-01398-2.pdf

Chapter 6

- González, L., & Méndez, C. (2024). Climate change and energy security: A comparative analysis of the role of energy policies in advancing environmental sustainability. Energies, 17(13), 3179. https://

- www.mdpi.com/1996-1073/17/13/3179
- Kleinman Energy Policy Center. (2023). Navigating the path to a sustainable energy future: Lessons from REFS 2023. Kleinman Center for Energy Policy. https://kleinmanenergy.upenn.edu/commentary/blog/navigating-the-path-to-a-sustainable-energy-future-lessons-from-refs-2023/
- World Economic Forum (WEF). (2020). Energy security and sustainable development.WorldEconomicForum. https://www.weforum.org/agenda/2023/01/davos23-energy-transition-security-sustainability-whitepaper/
- World Economic Forum (WEF). (2023). Making the future energy system secure meansmakingitsustainable.WorldEconomic https://www.weforum.org/agenda/2023/01/davos23-energy-transition-security-sustainability-whitepaper/

Chapter 7

- Deutch, J. (2008). The geopolitics of energy: From security to survival. Brookings Institution.https://www.brookings.edu/wpcontent/uploads/2016/07/energysecurity_chapter.pdf
- Mitchell, T., & Smith, A. (2023). The politics of energy security. Nature Energy, 8, 1096-1104. https://www.nature.com/articles/s41560-023-01398-2.pdf
- Munich Security Conference (MSC). (n.d.). Power shifts: On the geopolitics of theenergytransitionMunichSecurityConference. https://securityconference.org/en/news/full/power-shifts-geopolitics-energy-transition/
- Oxford Institute for Energy Studies. (2021). The geopolitics of energy: Out with

theold,inwiththenew. Oxford Energy Forum, (126), 1-15. https://www.oxfordenergy.org/wpcms/wp-content/uploads/2021/02/OEF-126.pdf

Chapter 8

- Accenture. (2022a). Energy security in energy transitions. https://www.accenture.com/us-en/insightsnew/energy/balancing-sustainability-security
- Accenture. (2022b). A balanced energy transition: Sustainability and security. https://www.accenture.com/us-en/insightsnew/energy/balancing-sustainability-security
- International Energy Agency. (2022). Energy security in energy transitions. In World Energy Outlook 2022. https://www.iea.org/reports/world-energy-outlook-2022/energy-security-in-energy-transitions
- Zhao, X., Zhang, L., & Yang, J. (2022). Are there conflicts among energy security, energy equity, and environmental sustainability in China's provinces? Sustainability, 14(11), 6873. https://www.mdpi.com/2071-1050/14/11/6873

Chapter 9

- Heal, G. (2013). The economics of energy security. National Bureau of Economic Research. https://www.nber.org/system/files/working_papers/w19729/w19729.pdf
- International Energy Agency. (2022). Energy security in energy transitions. In World Energy Outlook 2022. https://www.iea.org/reports/world-energy-outlook-2022/energy-security-in-energy-transitions
- Keefe, B. (2022). The price of energy insecurity. International Monetary Fund. https://www.imf.org/en/

- *Publications/fandd/issues/2022/12/POV-the-price-of-energy-insecurity-keefe*
- *World Economic Forum. (2022). Fostering effective energy transition 2022. https://www.weforum.org/publications/fostering-effective-energy-transition-2022/in-full/1-2-energy-security-and-access/*

Chapter 10

- *Energy Security in a Changing Geopolitical Landscape: Policy Challenges and StrategiesforEnergyMarketResilience. (2024).Bing. https://bing.com/search?q=Energy+Security+in+a+Changing+Global+Landscape*
- *International Energy Agency. (2024). Climate resilience for energy security. https://www.iea.org/reports/climate-resilience-for-energy-security*
- *International Energy Agency. (2024). Global energy perspective 2024. https://www.iea.org/reports*
- *International Energy Agency. (2024). Global energy perspective 2024. https://www.iea.org/reports/world-energy-outlook-2022/energy-security-in-energy-transitions*
- *Jaber, J. O. (2020). Energy security and renewable energy: A geopolitical perspective. In Renewable Energy (pp. 1-20). IntechOpen. https://www.intechopen.com/chapters/715*

Chapter 11

- *The Institute of World Politics. (n.d.). The role of rules and institutions in global energy: An introduction. https://www.iwp.edu/center-for-energy-security-and-diplomacy/*

- United Nations Economic Commission for Europe. (n.d.). The role of international organisations in enhancing energy security and managing environmental impacts. https://unece.org/fileadmin/DAM/energy/se/pp/EnCom16/28Nov07/1.1_Snoy.pdf
- United Nations Economic Commission for Europe. (2024). The energy security andclimateinitiative. https://unece.org/fileadmin/DAM/energy/se/pp/EnCom16/28Nov07/1.1_Snoy.pdf

Chapter 12

- Climate change and energy security: The dilemma or opportunity of the century? (2024). https://link.springer.com/article/10.1007/s10018-023-00391-z
- ClimateResilienceforEnergySecurity(2022). https://www.iea.org/reports/climate-resilience-for-energy-security
- ClimateResilienceforEnergySecurity(2022). https://www.oecd.org/en/publications/climate-resilience-for-energy-security_2a931f53-en.html
- Emerging Economies (2015). https://link.springer.com/book/10.1007/978-81-322-2101-2
- Unlocking renewable energy future in emerging markets (2024). https://www.weforum.org/impact/clean-energy-in-emerging-markets/

Chapter 13

- Building Trust through an Equitable and Inclusive Energy Transition (2024). https://www3.weforum.org/docs/WEF_Building_Trust_through_an_Equitable_and_Inclusive_Energy_Transition_2024.pdf
- FosteringEffectiveEnergyTransition(2022). https://

- www.weforum.org/publications/fostering-effective-energy-transition-2022/in-full/1-2-energy-security-and-access/
- *Making the future energy system secure means making it sustainable* (2023). https://www.weforum.org/agenda/2023/01/davos23-energy-transition-security-sustainability-whitepaper/

Chapter 14

- *Climate change and energy security: The dilemma or opportunity of the century?* (2024). https://link.springer.com/article/10.1007/s10018-023-00391-z
- *ClimateResilienceforEnergySecurity(2022).* https://www.iea.org/reports/climate-resilience-for-energy-security
- *ClimateResilienceforEnergySecurity(2022).* https://www.oecd.org/en/publications/climate-resilience-for-energy-security_2a931f53-en.html
- *The Adaptation Principles: 6 Ways to Build Resilience to Climate Change* (2020). https://www.worldbank.org/en/news/feature/2020/11/17/the-adaptation-principles-6-ways-to-build-resilience-to-climate-change

Chapter 15

- *Navigating the Path to a Sustainable Energy Future: Lessons from REFS 2023* (2023). https://kleinmanenergy.upenn.edu/commentary/blog/navigating-the-path-to-a-sustainable-energy-future-lessons-from-refs-2023/
- *Powering the Future: Risk Governance for a Sustainable, Resilient and Inclusive Energy System* (2024). https://www.undp.org/publications/

- *powering-future-risk-governance-sustainable-resilient-and-inclusive-energy-system*
- *Shaping a sustainable energy future in Asia and the Pacific: A greener, more resilientandinclusiveenergyfuture(2021). https://www.unescap.org/kp/2021/shaping-sustainable-energy-future-asia-and-pacific-greener-more-resilient-and-inclusive*
- *Sustainable Energy Transition for Renewable and Low Carbon Grid Electricity Generation and Supply (2022). https://www.frontiersin.org/journals/energy-research/articles/10.3389/fenrg.2021.743114/full*

ABOUT THE AUTHOR

Olugbenga Onungbogbo is a seasoned expert in the Energy field and has written several books on the subject 'From Crisis to Stability: Understanding Energy Security in a Volatile World is a book essential for policymakers, industry players, and anyone concerned about the future of our energy resources. It does not only highlight the challenges but also offers a roadmap for achieving a secure and resilient energy future.

www.ingramcontent.com/pod-product-compliance
Lightning Source LLC
Chambersburg PA
CBHW052335220526
45472CB00001B/437